Optimization and Mathematical Modeling in Computer Architecture

Synthesis Lectures on Computer Architecture

Editor
Mark D. Hill, *University of Wisconsin, Madison*

Synthesis Lectures on Computer Architecture publishes 50- to 100-page publications on topics pertaining to the science and art of designing, analyzing, selecting and interconnecting hardware components to create computers that meet functional, performance and cost goals. The scope will largely follow the purview of premier computer architecture conferences, such as ISCA, HPCA, MICRO, and ASPLOS.

The Memory System: You Can't Avoid It, You Can't Ignore It, You Can't Fake It
Bruce Jacob
2009

Fault Tolerant Computer Architecture
Daniel J. Sorin
2009

The Datacenter as a Computer: An Introduction to the Design of Warehouse-Scale
Machines
Luiz André Barroso and Urs Hölzle
2009

Computer Architecture Techniques for Power-Efficiency
Stefanos Kaxiras and Margaret Martonosi
2008

Chip Multiprocessor Architecture: Techniques to Improve Throughput and Latency
Kunle Olukotun, Lance Hammond, and James Laudon
2007

Transactional Memory
James R. Larus and Ravi Rajwar
2006

Quantum Computing for Computer Architects
Tzvetan S. Metodi and Frederic T. Chong
2006

Optimization and Mathematical Modeling in Computer Architecture

Tony Nowatzki, Michael Ferris, Karthikeyan Sankaralingam, Cristian Estan, Nilay Vaish, and David Wood

ISBN: 978-3-031-00645-6   paperback
ISBN: 978-3-031-01773-5   ebook

DOI 10.1007/978-3-031-01773-5

A Publication in the Springer series
SYNTHESIS LECTURES ON COMPUTER ARCHITECTURE

Lecture #26
Series Editor: Mark D. Hill, *University of Wisconsin, Madison*
Series ISSN
Synthesis Lectures on Computer Architecture
Print 1935-3235   Electronic 1935-3243

Optimization and Mathematical Modeling in Computer Architecture

Tony Nowatzki
University of Wisconsin-Madison

Michael Ferris
University of Wisconsin-Madison

Karthikeyan Sankaralingam
University of Wisconsin-Madison

Cristian Estan
Broadcom Corporation

Nilay Vaish
University of Wisconsin-Madison

David Wood
University of Wisconsin-Madison

SYNTHESIS LECTURES ON COMPUTER ARCHITECTURE #26

ABSTRACT

In the last few decades computer systems and the underlying hardware have steadily become larger and more complex. The need to increase their efficiency through architectural innovation has not abated, but quantitatively evaluating the effect of various choices has become more difficult. Performance and resource consumption are determined by complex interactions between many modules, each with many possible alternative implementations. We need powerful computer programs to explore large design spaces, but the traditional approach of developing simulators, building prototypes, or writing heuristic-based algorithms in traditional programming languages is often tedious and slow.

Fortunately mathematical optimization has made great advances in theory, and many fast commercial and academic solvers are now available. In this book we motivate and describe the use of mathematical modeling, specifically optimization based on mixed integer linear programming (MILP) as a way to design and evaluate computer systems. The major advantage is that the architect or system software writer only needs to describe what the problem is, not how to find a good solution. This greatly speeds up their work and, as our case studies show, it can often lead to better solutions than the traditional approach.

In this book we give an overview of modeling techniques used to describe computer systems to mathematical optimization tools. We give a brief introduction to various classes of mathematical optimization frameworks with special focus on mixed integer linear programming which provides a good balance between solver time and expressiveness. We present four detailed case studies—instruction set customization, data center resource management, spatial architecture scheduling, and resource allocation in tiled architectures—showing how MILP can be used and quantifying by how much it outperforms traditional design exploration techniques. This book should help a skilled systems designer to learn techniques for using MILP in their problems, and the skilled optimization expert to understand the types of computer systems problems that MILP can be applied to.

Fully operational source code for the examples used in this book is provided through the NEOS System at http://www.neos-guide.org/content/computer-architecture

KEYWORDS

Integer Linear Programming, ILP, Mixed Integer Linear Programming, MILP, Mathematical Modeling, General Algebraic Modeling System, GAMS, Optimization, Spatial Architectures, Tiled Architectures, Scheduling, Resource Allocation, Instruction Set Customization

Contents

---

[1]★ indicates an optional section.

Acknowledgments

We would like to acknowledge and thank many of you who contributed to this synthesis lecture. Thanks to Mark Hill for helping refine the scope of the book, constant encouragement, and reviewing drafts of this lecture. Thanks to Benjamin Lee, Lieven Eeckhout, Paul Feautrier, and other anonymous reviewers for several comments that helped improve this lecture. Thanks to Newsha Ardalani, Michael Bussieck, Preston Briggs, Daniel Luchuap, Zach Marzec, and Lorenzo De Carli for reading drafts in detail and providing feedback. Thanks to Somesh Jha for his help on early work in applying SMT techniques to spatial architecture scheduling. Thanks from Nilay to Srikrishna Sridhar and Taedong Kim for the numerous discussions on mathematical optimization.

Thanks to Michael Sartin-Tarm for creating the online case studies for this synthesis lecture.

Much of the content of this lecture is built on research that has been supported by various grants from the Air Force Office of Scientific Research, the Department of Energy and the National Science Foundation. We are grateful for their vision to facilitate both focused disciplinary research and the development of new areas, applications, and impacts arising from interdisciplinary interactions.

Tony Nowatzki, Michael Ferris, Karthikeyan Sankaralingam, Cristian Estan, Nilay Vaish, and David Wood
September 2013

CHAPTER 1

Introduction

1.1 WHY THIS BOOK?

The past half century has seen an important culmination of trends. First, the mathematical theory behind optimization has evolved to a state where it can be considered mature, and this has spurred classes of entirely new and significantly more powerful methods for solving optimization problems. Second, the robustness, capabilities, and performance of academic and commercial optimization solvers, along with new ways for expressing real world systems using mathematical constructs and languages, has greatly matured and shows no sign of slowing progress. Third, architecture and systems have become more complex, necessitating sophisticated analysis for modeling and advanced algorithms for design. The power of mathematical tools enable reasoned analysis of these complex systems. Due to these trends, the use of optimization for computer systems in general and computer architecture specifically becomes ever more practical and helpful, and therefore continues to see increased use. These broad trends are outlined below.

1.1.1 EVOLUTION OF MATHEMATICAL THEORIES AND ALGORITHMS

In 1947, George Dantzig invented the simplex algorithm for linear programming,[1] one of the fundamental building blocks of mathematical optimization. Since then, the field of applied mathematical optimization has burgeoned extensively; for example, the 1975 Nobel prize in Economics was awarded to Kantorovich and Koopmans for "their contributions to the theory of optimal allocation of resources." The 1979 paper of Khachiyan [105], proving that linear programming was solvable by a polynomial-time algorithm, led to a huge resurgence in research of efficient methods for large-scale problems, and the 1984 paper of Karmarkar [102] demonstrated that an interior point algorithm, an alternative to the simplex method, could in fact be both theoretically powerful, and practically efficient. Currently, both the simplex method and interior point methods are used in large scale settings due to their different theoretical and practical behaviors. The foundation for the theory of duality is widely credited to von Neumann [173], and the dual simplex method [117] that is a critical component of modern day Mixed Integer Linear Programming (MILP) codes was described by Lemke in 1954. Network flow problems were of great interest at the outset due to their practical significance, and their special structure has led to many important enhancements of linear programming theory, coupled with specialized algorithms to exploit their structure [7]. The obituary article of Dantzig [41] gives a more complete timeline of the developments of linear programming. Besides models for purely linear systems, there are opti-

---

[1]Mathematical terminology defined in Chapter 2.

mization techniques for integer variables, logical constraints, quadratic relationships, the modeling of uncertainty, and even general nonlinear and non-smooth systems. For example, the Kuhn Tucker conditions [111] established a theoretical foundation for nonlinear programming in the early 1950s, and that field developed to encompass methods for regression and parameter estimation in statistics, optimal control and engineering design. Duality theory was extended to the broader class of convex analysis, neatly summarized in 1970 by Rockafellar [149]. Extensions of this theory include semidefinite optimization [22], which is a very expressive theoretical tool with powerful algorithm properties. This has led to a huge improvement in solver technologies over the past decade.

The field of combinatorial optimization, starting from the contributions of Gomory [78], has also developed rapidly and has allowed deep and important insights into the underlying properties of optimization models including discrete decisions [139, 152, 179]. The last two decades have also seen an explosion in the use of these discrete techniques in practical problems, driven in part by the improvements in large scale linear programming codes and the theory and implementation of cutting planes. The huge improvement in computational resources has also led to increased research in the area of stochastic programming—the theory in this area nicely joins together optimization and probability [155]. The interplay between optimization and statistics is also at the heart of much research in sparse optimization or compressed sensing, and this has led to increased interest in first-order methods that can be applied to problems with huge datasets.

1.1.2 MATURITY OF SOLVERS AND MODELING SYSTEMS

The demand for high-performance optimizers, coming from industrial systems optimization, as well as economic applications has driven the field from its outset through the present day. The first commercial MILP code was developed by Beale and colleagues [19] in 1965, and used on a number of industrial applications (see [28] for further details). Various other commercial codes were developed by leading computer, oil, and independent optimization specialist companies. Academia also played a central role with codes such as MINOS [136], paving the way for open source counterpart development and innovation.

Since the addition of MILP solving capabilities to CPLEX 1.2 in 1991, the total accumulated *machine independent* speedup factor up to CPLEX 11 is nearly 30,000 times [28]. Combined with the speedups from device scaling and computer architecture itself, many problems that were once wholly intractable are now quite practical.

Optimization modeling systems are a way to bring the techniques of computational optimization to powerful applications, and their development since the 1970s has led to a new period of extensive use of optimization. The surveys given in [69, 112] outline the path from matrix generators to modern day modeling systems, while [100] details recent enhancements that are intended to help solve real world problems. As the field grows in generality and robustness, the types of applicable problems continues to increase, and impact the fields of business, economics, engineering, medicine, and science. Survey papers [65, 107, 156] and handbooks dedicated to the use of optimization in specific domains abound, e.g., [145].

1.1.3 COMPLEXITY OF COMPUTER SYSTEMS

Due to the increased complexity of computer systems, mathematical optimization has pervaded the field of computer hardware design for at least the last three decades. This spans the range of optimization in VLSI design automation, architecture examples like automatic generation of accelerators and Network-on-Chip optimization. Also, mathematical optimization is extremely important for certain compiler analysis, like data-dependence analysis, and this related work is revisited in Chapter 5. To motivate the case for this book, we briefly outline the research in VLSI design and computer architecture of the past and present.

VLSI   Perhaps the most prominent applications of optimization techniques have been in the field of VLSI. Indeed, almost all levels of VLSI physical design automation have been solved with Mixed Integer Linear Programming. The floorplanning problem, where hardware modules are placed on a 2D plane to minimize the total area, are a natural fit for MILP. It has seen much research throughout the 1990s, and into the 2000s, and prominent works are by Sutanthavibul et al., Sen et al., and Dorneich et al. [51, 55, 154, 160]. Global routing, where the approximate connections between blocks are determined, are also a good fit for MILP. Usually since the scale of these algorithms are large, relaxations and approximations of the MILP are used to solve the problem [20, 165, 180]. Not only has MILP been applied to the problems in VLSI design automation, but also in the related field of scheduling tests for verification [36, 37].

Integer linear programming sees continued use and research in VLSI technologies, including floorplanning with module selection in mixed granularity FPGAs [157], and floorplanning in 3D manufacturing to reduce the number of 3D vias subject to area and power constraints [97]. Also, MILP has been recently applied to routing for flip-chip interconnects [62]. An excellent reference on optimization in this field is "Combinatorial Optimization in VLSI Design" by Held et al. [91].

Computer Architecture   To provide a contrast between VLSI and computer architecture problems, we characterize them as follows. VLSI problems have well-defined objective functions, working on circuits already represented as a graph, and the primitive element is a gate transistor or even lower. There are millions of nodes/nets or other variables. Architecture problems tend to involve generally coarser grained blocks, like functional units or processing elements, interacting in an ad-hoc, or more unrestricted fashion. Problems are not already formulated with a graph representation, and the objective function can be unclear. The coarse grain nature leads to many fewer decisions, generally in the range of 10s to 100s. We discuss some examples next.

In the accelerator domain, optimization has seen a variety of uses. Lee et al. use integer linear programming for scheduling the execution of hardware datapaths, subject to timing and resource constraints [115]. Similarly, Azhar et al. use MILP for specializing the datapath for a

Viterbi decoding accelerator [13]. An important field of research focuses on the interconnect for SOC designs, where the solution is to employ a network on chip. The interconnect topology and bandwidth can be specialized to an application to avoid over provisioning. Srinivasan et al. solve this problem using a two stage MILP formulation, minimizing the power consumption subject to power constraints [159]. The performance of systems can be dramatically affected by which pieces are performed in hardware or software. The trade-off in placing computation on the hardware is largely to improve performance, but its costs involve area and power. One important work in this field is by Niemann et al. [140], who take a VHDL specification of an application, and use integer linear programming to separate the specification into software and hardware components.

In this book, we study four concrete use cases for MILP in the field of computer architecture, chosen to demonstrate both MILP's range of applicability and expressive power in terms of modeling interesting system features.

1.2   WHO IS THIS BOOK FOR?

Before getting into an overview of the contents of this book, we briefly discuss the dual intended audiences of this book.

Systems perspective for the skilled optimization expert   This book provides a systems perspective on the nature of problems encountered in computer systems by examining in detail four case studies in which mathematical modeling is applied to the design and evaluation of four very different systems. Thus it gives the optimization expert background and perspective on systems problems, helping them approach and solve such problems using their insight and expertise.

Modeling techniques for the systems expert   This book provides a general overview of mathematical modeling and optimization specifically for the systems expert interested in learning about such techniques. Specifically we use Mixed Integer Linear Programming, covering its basic theory and practical implementation and uses. We discuss how MILP modeling is used to design and evaluate four diverse types of systems. The specific system aspects considered in this book are: instruction set customization for a processor, data center job scheduling, and compiler/microarchitecture design of spatial computing architectures.

1.3   WHAT IS THIS BOOK ABOUT?

Driven by the above trends, we predict that mathematical modeling will play an increasing role in the design and evaluation of future computer systems. This book is about using modeling to design and evaluate computer systems, specifically an optimization technique called *mixed integer linear programming* to design and evaluate *systems commonly encountered in computer architecture*. To provide context to what this book is about, we briefly review what mathematical modeling is and then describe optimization, which is a type of modeling. We then describe the essential primitives that make up mixed integer linear programming, and discuss why this is suitable through two examples: i) a simple intuitive example that should be easily accessible to the reader, and ii) a somewhat non-intuitive problem that does not at first glance fit the primitives of logic constraints that make up MILP. The second example is chosen to illustrate the power of MILP.

1.3.1 MATHEMATICAL MODELING

Modeling is a descriptive process: it validates principles and/or explores underlying mechanisms. Modeling complements the design and evaluation of systems using simulation or prototype building. Modeling typically involves repeated solutions to gain an "understanding" of the solution space (or sensitivity). It is always important to demonstrate models on a problem that is understood by the audience and we intend to do that in this volume. This can lead to difficulties. For example, how much detail of the underlying system can we expose to the mathematical model, i.e., is it a black box system, or are some details available to the model—a so-called white box system? There is also lots of subjective information, which if not correctly handled leads to unnecessary complexity. Models must be trained and continually evaluated.

A mathematical model is the description of a system using mathematical concepts and language. Such models are used in the natural sciences (such as physics, biology, earth science, meteorology), engineering disciplines (e.g., computer science, electrical, mechanical and chemical engineering), and in the social sciences (such as economics, psychology, sociology and political science). Physicists, engineers, statisticians, operations research analysts, and economists use mathematical models extensively to investigate underlying physical phenomena, to design new systems and to enhance the operation of existing systems. A lack of agreement between theoretical mathematical models and experimental measurements often leads to important advances as better theories are developed.

In many systems engineering problems, the underlying system is described by a collection of linear ($Ax = b$) or nonlinear equations ($F(x) = 0$) involving "state" variables x, whose values define the underlying system and how it operates in steady-state (and possibly describe its evolution over time). Theoretical investigations consider when solutions exist, when they are unique, how stable they are to perturbations in the defining data, and how to efficiently find values that satisfy the equations.

Modeling systems must be able to model the problem at hand in an easy and natural manner and can be built in a number of very different ways. These range from pencil and paper scrawlings, to Excel, Matlab, or R models, or more sophisticated implementations using scripting or programming languages, or specialized modeling settings. In computer architecture and systems, mechanistic models based on average mean value analysis and queuing theory are commonly used [53]. Our focus here is on a particular type of mathematical model, an optimization problem, that involves not only equations and variables, but also the concepts of inequality constraints, discrete variables, and objective functions. These concepts are defined and manipulated by the growing number of modeling systems for optimization, including GAMS, AMPL, AIMMS, and others as we detail later in Chapter 2.

1.3.2 OPTIMIZATION AS A MODELING TECHNIQUE

Optimization is a specific type of mathematical model that facilitates the improvement or design of an underlying system. We are particularly interested in models that can be used for design. Such models can investigate certain properties of new designs, including whether they can be economical given the price of raw materials, production processes, etc., or to consider various alternatives when ascertaining a strategic plan. We discuss two specific examples in computer architecture shortly. To use optimization for design, a modeler must understand the key driving mechanisms within the design and be able to construct a simplification or (mathematical) abstraction of the true system. Often, just solving a single problem isn't the real value of modeling: optimization finds "holes" in the model that enable, through an iterative process, a richer and more complete/accurate representation of the system at hand.

Many optimization models have been developed, each of which aptly models certain situations and environments. These models have varying levels of expressive power, and in turn can be solved more or less efficiently. By understanding the principles underlying optimization models and the relationships between them, it becomes much simpler to choose the correct model for the correct task. Fortunately, modeling languages, which express optimization problems cleanly in terms of their variables, constraints, and objectives, can usually be used for a variety of optimization models. This makes it easy to move between optimization models as the requirements of the system or problem changes. Chapter 2 describes optimization models, as well as solvers and modeling languages, in more detail.

The modeling system that we use, both in the examples and case studies, is GAMS (General Algebraic Modeling System). This is just one possible modeling language, and the modeling techniques we discuss could be applied with any of the other modeling languages. Also, we focus specifically on Mixed Integer Linear Programming (MILP) because of its expressive power, declarative nature, and the presence of many fast commercial/free solvers. Another benefit of MILP, as with many other optimization models, is that problems can be either solved to optimality, or solved until they reach some bound on the optimality gap. We describe next the essential primitives of MILP.

1.3.3 THE ESSENTIAL PRIMITIVES OF MILP

The following primitives capture the essence of what it means to model a design or system exploration problem as a MILP. Following this description we explain two problem formulations using the intuitive concepts and then using mathematical notation.

1. All system features that have freedom to be changed by the designer are expressed as *variables*.

2. All system features that are fixed, we refer to as *parameters*; they essentially become constants in mathematical formulas.

3. The behavior of the system is expressed using a collection of *functions* of the variables and parameters.

4. All system features that restrict its behavior are expressed as *constraints* on these functions.

5. The system property that the designer wants to optimize is selected from these functions and is termed the *objective function*.

6. The objective function and the collection of constraints together form the *model* of the system.

7. For MILP, some variables can take only integer values and constraints are linear relationships of variables.

Based on the range and types of values variables can take, the relationship between variables that make up constraints, and the nature of the final objective(s), there are various "families" of optimization models including linear programming, convex quadratic programming, etc. These relationships are developed in detail in Chapter 2. This book and its examples focus on MILP, because it provides a nice balance between expressive power, solution time, and ability to provide optimality guarantees.

1.3.4 ILLUSTRATIVE EXAMPLES

To demonstrate both the intuitive nature of MILP, and its perhaps surprising expressive power, we give two example problems and solutions. For reference, following the mathematical descriptions are the corresponding GAMS programs, in Figures 1.1 and 1.2. This code can be used directly without modification.

Example 1: Special Function Units   A processor architect must choose some number of special functional units (SFUs) based upon the cost and projected performance benefits. SFUs are specialized pieces of hardware, like a sin/cos unit or a video decoder, which are good at specific computations or tasks. Specifically, there are N special function units which improve the performance of a specific task, each one requiring certain chip area, A_n. The maximum area budget is MAX_{area}. Also, there are M applications, and it is known how much each special function unit can speedup each application in M, called S_{mn}. The values in A_n, MAX_{area} and S_{mn} are the input parameters of the system. The architect must choose which special function units to operate or implement, represented by O_n. Here, $\{O_n \mid n = 1, \ldots, N\}$ is the set of binary variables which describe the choices in our system. The performance improvement on an application $m \in M$ is $\sum_{n=1}^{N} S_{mn} * O_n$, where each improvement S_{mn} is only counted when unit n is operational, i.e., $O_n = 1$. If the goal is to improve all applications' performance by some minimum amount, the following objective function and constraints form the model of interest:

$$\max_{O} \quad \min_{m=1,\dots,M} \sum_{n=1}^{N} S_{mn} * O_n$$

$$\text{s.t.} \quad \sum_{n=1}^{N} A_n * O_n \leq MAX_{area}$$

$$O_n \in \{0, 1\} \qquad n = 1, \dots, N$$

In the notation above, "$\max_{O}$" indicates that members of O are variable in the following problem. The expression to the right is what we are maximizing, which is the minimum performance improvement over all applications. The first constraint above, which appears after the "s.t.", restricts the total area to the maximum, and the second constraint ensures that all variables in O are binary.

As we outline in Section 2.2.2 (or by elementary observations), this model can be recast as a MILP, using an additional variable $PERF$ to model the lower-bound performance improvement of all applications:

$$\max_{O, PERF} \quad PERF$$

$$\text{s.t.} \quad \sum_{n=1}^{N} A_n * O_n \leq MAX_{area}$$

$$PERF \leq \sum_{n=1}^{N} S_{mn} * O_n \quad \text{for all } m \in M$$

$$O_n \in \{0, 1\} \qquad n = 1, \dots, N, PERF \in \mathbb{R}$$

In this case, the objective function is simply to maximize $PERF$. The first constraint above remains the same, and the second constraint calculates the performance gain based on which SFUs are selected, and limits $PERF$ to the minimum performance gain across applications.

Example 2: Instruction Scheduling   Instruction scheduling is a compiler optimization which orders instructions to achieve a maximum amount of instruction level parallelism. Here, we consider the problem of scheduling a basic block for a multi-issue in-order processor (with issue width r). The number of instructions in a basic block is N. The set D_{ij} lists pairs of data-dependent instructions, i.e., pairs such that instructions j dependents on the output of instruction i. The expected latencies between dependent instructions are captured by L_{ij}. The values of D_{ij} and L_{ij} are the systems input parameters. The instructions are mapped into particular clock cycles, the maximum number we consider is C_{max}. The schedule is represented by the set of binary variables $cycle_{ic}$, which indicate if instruction i is mapped into cycle c. The objective, in this example, is to minimize the total cycles necessary (the latency), described by the variable LAT. Good modeling practice also introduces an additional variable $number_i$ that describes the cycle number that instruction i executes on. Adding variables that have physical meaning often can lead to tighter

```
$set N 50                         | M and N are pre-processor global variables
$set M 30                         | indicating the problem size.
                                  |
                                  |
* SFUs                            | Here, we create N SFUs and M applications,
Set n /n1*n%N%/;                  | using the %var% syntax to access global
* Applications                    | variables.
Set m /m1*m%M%/;                  |
                                  |
* Parameters                      | These are the problem parameter definitions.
Parameter S(m,n);                 | The are defined over the previous sets, m for
Parameter A(n);                   | applications, and n for SFUs.
Scalar MAXarea;                   |
                                  |
* Define params randomly for this example. | GAMS provides built-in math functions like
S(m,n)=uniform(0,1/%N%);          | max and also uniform/normal for generating
A(n)=max(normal(1,.5),0);         | random numbers. These calculations occur
MAXarea=%N%/5;                    | before model generation/evaluation.
                                  |
* Define variables.               | Here the variables O_n and PERF are
Binary Variable O(N);             | declared.
Variable PERF;                    |
                                  | Our two equations are declared and defined.
* Declare equations.              | The first set of parens after the equation
Equations limArea,limPerf(m);     | name indicates the set over which to cre-
                                  | ate equations. For example, limPerf(m),
* Define equations.               | defines M equations, one for each appli-
limArea..     sum(n,A(n)*O(n)) =l= MAXarea; | cation. Similarly, sum(n,expr) means to
limPerf(m).. PERF =l= sum(n,O(n)*S(m,n)); | take the sum of expr the expression over set n.
                                  |
* Declare and solve MILP model.   | Finally, we create the model and solve us-
Model sample1 /limPerf,limArea/;  | ing MIP (Mixed Integer linear Programming),
solve sample1 using MIP maximizing PERF; | minimizing the objective variable PERF.
                                  |
```

Figure 1.1: GAMS Code Example 1 (Special Function Unit Selection).

models. It is important for clarity and efficiency to introduce them if they appear multiple times within the model. The following describes the above model:

$$\min_{cycle,number,LAT} LAT$$

$$\text{s.t.} \quad \sum_{c=1}^{C_{max}} cycle_{ic} = 1 \quad \text{for all i} \in \text{N}$$

$$\sum_{i \in N} cycle_{ic} \leq r \quad \text{for c from 1 to } C_{max}$$

$$number_i = \sum_{c=1}^{C_{max}} c * cycle_{ic} \qquad \text{for all i} \in \text{N}$$

$$number_i + L_{ij} \leq number_j \qquad \text{for i,j} \in D_{ij}$$

$$LAT \geq number_i \qquad \text{for all i} \in \text{N}$$

$$cycle_{ic} \in \{0, 1\}, LAT \in \mathbb{R}, number_i \in \mathbb{R}$$

The first equation enforces that all instructions are be mapped to one of the clock cycles. The second constraint enforces the issue width of the processor. The expression $\sum_{c=1}^{C_{max}} c * cycle_{ic}$ describes the cycle number that instruction i executes on. The fourth equation is the most complex, but simply ensures that dependent instructions are scheduled at least the required number of cycles apart. The final constraint computes the objective, the latency of the basic block, by constraining LAT to be larger than any of the individual instructions' latencies. Minimizing LAT is the objective function.

1.3.5   BENEFITS OF MODELING AND MILP

Compared to system design and optimization using simulation and building prototypes, modeling in general and MILP specifically provides many benefits, which are:

- First, compared to simulation and building systems, modeling is much faster. In computer architecture for example, modeling is several orders of magnitude faster than studying systems with simulation. Essentially MILP solvers can perform fast enumeration and pruning of possibilities.

- Modeling expresses the design problem in a declarative fashion, which contrasts with implementing/simulating a system, performed imperatively or heuristically. For the problems we study in this book, heuristic-based techniques are typically written using many lines of imperative code like C/C++ which implement a heuristic to solve the problem. In contrast, a MILP program can declaratively specify the original problem to be solved in a very concise fashion. For reference, Table 1.1 shows the lines of GAMS code required for each of the models in the case studies in this book, ranging from 20 to 200 lines. This is significantly less than the 1,000s to 10,000s of lines of C++ or Java code required for some heuristic-based versions.

- Modeling can often provide deep insight on the problem and "holes" in the design. Sometimes these are just omissions to the model formulation, but often they reveal new mechanisms that the modeler was unaware of. In many cases, an optimal solution reveals an

```$setglobal N 100```	Global N defines the number of instructions.

```Set n /n1*n%N%/;```
```set c /1*%N%/;```
```alias (n,i,j);```
	The alias command creates new names for the same set.

```* Dependences and Latency from i to j```
```Set D(i,j);```
```Parameter L(i,j);```
	D(i,j) is a multi-dimensional set over the aliased set n. Equations which involve D will use the aliased i and j to distinguish between the first and second index.

```* Generate Random DAG```
```D(i,j) = yes $(ORD(j)>ORD(i)```
```   and uniform(0,1)>0.8);```
```L(D) = round(uniform(0,3));```
```scalar r/2/;```
	We generate a random instruction dag as the problem's input. The dollar sign $ is a conditional operator. Here, D(i,j) is included in the set (=yes) if the dollar sign expression is true. Also, = ORD(i) is i's rank inside set n.

```binary variable cycle(i,c);```
```positive variables number(i);```
```variable LAT;```
	Variable Declarations

```Equations unique(i),width(c),```
```   def_number(i),deps(i,j),obj(i);```
```unique(i)..   sum(c,cycle(i,c))=e=1;```
```width(c)..   sum(i,cycle(i,c))=l=r;```
```def_number(i)..```
```   number(i) =e= sum(c,ORD(c)*cycle(i,c));```
```deps(i,j)$D(i,j)..```
```   number(i) + L(i,j) =l= number(j);```
```obj(i)..   LAT =g= number(i);```
	Declarations and definitions of the equations. Note that model equations (as opposed to calculations not in the model), use operators like "=e=","=l=", and "=g=" to represent $=$, $\leq$, and $\geq$ respectively.
	Also, the deps equation shows how the dollar sign can be used to generate constraints given a certain condition. Here, deps is only generated if j depends on i.

```Model sample2b/all/;``` ```solve sample2b using MIP minimizing LAT;```	Define the problem using all equations, and solve using MIP while minimizing LAT.

Figure 1.2: GAMS Code for Example 2 (Instruction Scheduling).

implicit constraint, or determines a result that is unacceptable. These can be corrected by changing the model data and constraints. Sometimes it reveals a more fundamental issue that invalidates part of the model description or shows a particular mechanism can function in a way that is different to that required. This may require serious structural changes to the model functions and variables, and sometimes points to the necessity of a completely new model. The process of learning often provides more value than the actual solution numbers themselves. In our experience, we found that applying MILP to solve spatial architecture scheduling revealed an important missing "hole" in previous approaches—in that they do not explicitly model or consider balancing throughput of computation with latency.

Modeling practitioners should weight the benefits of MILP along with the trade-offs and limitations. For instance, certain optimization problems may not be suitable for MILP because they are too large, non-linear, or inherently dynamic. In Chapter 7, we discuss the characteristics

Case Study	GAMS Lines of Code
1. Instruction Set Customization	73
2. Data Center Resource Management	22-53
3. Spatial Architecture Scheduling	20-41
4. Resource Allocation in Tiled Architectures	198-239

Table 1.1: Lines of code for case study models (does not include inputs or solver flags).

of "MILP-friendly" problems, along with practical strategies for deciding when to use MILP and how to speed up MILP formulations.

1.4   WHAT THIS BOOK IS NOT ABOUT

This book provides background on modeling and optimization in Chapter 2 and discusses four case studies across various domains in successive chapters. This book is not a broad reference on modeling—we barely scratch the surface. It does not attempt to compare and contrast different modeling approaches, since this is a time consuming and risky endeavor. Neither is it a reference on all optimization approaches; it tackles *one* important class—MILP. Our conclusion chapter provides guidance on when MILP is well suited to a design problem, which the reader can use to guide their choice of modeling approach. For computer systems designers used to the parlance of mechanistic models and empirical models, our MILP-based modeling can be thought of as a hybrid.

From the optimization expert's perspective this book should *not* be viewed as a broad reference on computer systems problems. We cover four case studies across different domains to give a flavor of the nature of problems. Many problems in these domains have much larger variable sets, simpler or more complex relationships between variables, and fall within different optimization model families. We hope to give the optimization expert an overview of the types of problems to motivate them to apply their expertise in computer systems design.

1.5   BOOK OVERVIEW

We briefly describe the organization of this book below. Note that case studies 1 and 2 give introductory examples to modeling real-world computer-architecture problems, focusing on MILP modeling techniques, while Chapters 3 and 4 describe complex formulations and provide in-depth evaluations, focusing on comparing mathematical to heuristic methods, and using advanced techniques. The detailed overview is as follows:

- **Overview of Optimization** In this chapter we give an overview of mathematical modeling and various optimization theories. We describe mathematical modeling abstractly, then give an overview of important mathematical optimization methods. We subsequently describe how to model common phenomena in Mixed Integer Linear Programming (MILP) and offer some insight on solving techniques.

- **Case Study 1: Instruction Set Generation** In the domain of application-specific processors, generating custom instruction candidates from an application is the process of analyzing the intermediate representation of the program, and finding subgraphs which can be made into candidate instructions. The optimization problem we solve is to determine the convex set of operations inside a basic block, forming a custom instruction template, which maximizes the expected latency reduction.

- **Case Study 2: Data-Center Service Allocation** Data-center resource managers are centralized software components which manage the execution of a large number of services. Since co-locating multiple services on the machine could degrade performance, it is critical that the resource manager effectively allocates machine resources. Here, the optimization problem is to statically determine the best co-locations of services on servers such that the resource requirements and service level agreements can be satisfied.

- **Case Study 3: Spatial Architecture Scheduling** Hardware specialization has emerged as an important architectural approach to sustaining performance improvements of microprocessors despite transistor energy efficiency challenges. Specialization removes inefficiencies of traditional general purpose processing. One specialization strategy is to "map" large regions of computation to the hardware, breaking away from instruction-by-instruction pipelined execution and instead adopting a *spatial architecture* paradigm. The optimization problem we solve is to schedule computation nodes (in a Directed Acyclic Graph) to the spatial architecture's ISA-exposed hardware resources.

- **Case Study 4: Resource Allocation in Tiled Architectures** Many-core chip multiprocessors face complex resource allocation problems. For instance, they require multiple memory controllers to provide DRAM bandwidth, and must decide which tiles they should be placed on. This placement affects best-case latency and contention for network channels and routers. Also, the network fabric itself can be heterogeneous, and it must be decided where to place resources like buffers and virtual channels to minimize the congestion. This case study explores the combined memory controller placement problem and the heterogeneous network dimensioning problem.

1.6 CODE PROVIDED WITH THIS BOOK

Fully operational source code for the examples used in this book is provided through the NEOS system. The NEOS system (www.neos-server.org) is an online resource for optimization that provides access to various solvers and many more case studies [43] across various domains beyond what is discussed in this volume. It is a general system for optimization that handles broader classes of problems, such as nonlinear programming (NLP) [141], stochastic programming [155], complementarity or equilibrium problems (CP) [42, 60], hierarchical modeling, second order cone programs (SOCP) [9, 119] and semidefinite programs (SDP) [169], among many others. While these models have different expressive power, and are useful in many application domains, we believe that many interesting design optimization problems in computer architecture fit nicely into the more confined setting of mixed integer linear programming, and our purpose here is to describe these clearly and succinctly.

On the "Morgan Claypool Synthesis Lecture" web page on the NEOS server, we have provided the following:

- Source code for all four case studies, implemented using the GAMS modeling system. This code can be run on any machine that has a GAMS license.

- Instructions on running the source using the NEOS server, which does not require an end user to have their own GAMS license etc.

- An Internet-based applet and interactive demo, that allows users to modify and transform the models in place and run in the NEOS server.

Please find our NEOS webpage at:

http://www.neos-guide.org/content/computer-architecture

CHAPTER 2

An Overview of Optimization

This chapter serves two broad purposes. First, it provides a general overview of optimization techniques. Second, it provides a detailed treatment of how to model design problems as MILP including the selection and formulation of variables, constraints, objective functions, and some advanced techniques to reformulate nonlinearity into linearity. For those who want to quickly learn the basic material required to understand the book's case studies and quickly learn the techniques of MILP, we recommend reading sections 2.1, 2.2.1, 2.2.4, and 2.3, while less essential sections are marked with a star ★. Those who want a broader understanding of the field of optimization should read the entire chapter.

The chapter is organized as follows. We provide a general overview of optimization in the next section, followed by three sections that contain an introduction to optimization from a mathematical perspective. First, we give an overview of selected mathematical optimization models. The subsequent section describes how to model interesting and useful phenomena in Mixed Integer Linear Programming (MILP), and we conclude by giving some insights into the mechanisms of MILP solvers.

2.1 OVERVIEW OF OPTIMIZATION

In the most basic sense, optimization models consist of three components. First is the set of *variables*, which are the unknown quantities in the problem. The values of these will describe to the modeler the chosen design (or optimal solution). Next, the *domain* defines the allowable values that the variables can take on, through expressing relationships between variables. We often refer to these relationships as constraints. In design problems, these constraints describe aspects of the system which either restrict the solutions or attempt to quantitatively capture some behavior of the system. Finally, the *objective function* defines the desired property that a choice of the model's variables should improve. The modeler's optimization goals are encoded in this objective function.

Definitions   To aid the discussion of different types of optimization models, we briefly describe some terms:

Continuous variables Variables which can take on arbitrary real values.

Integer variables Variables which can take on only discrete, integer values. Binary variables are a specialization that can only take the values 0 or 1.

Linear constraints Equations or inequalities which allow only linear relationships between variables. In practice, this means variables can only be multiplied by constants and summed.

Nonlinear constraints Equations or inequalities which allow arbitrary (but typically differentiable) relationships between variables.

Linear objective A function defining the optimization goal, which is expressed in terms of linear relationships between model variables.

Convex domain The set of allowable values for a model, such that a line segment between any two points in the set is completely contained inside the set.

Convex objective An optimization function whose epigraph (the set of points on or above the function's graph) is a convex set. All local extrema of a convex function over a convex domain are global extrema.

Network representation A special representation of graphs, based on node-arc incidence, leveraged to improve the efficiency of network modeling.

Depending on the types of variables, domains, and objective functions which are allowed, different optimization models are created. One basic model we consider is Linear Programming (LP), where variables must be continuous, and all the relationships between variables and the objective function must be linear. The most complex model we mention is Mixed Integer Nonlinear Programming (MINLP), where some variables can be forced to be integral, and the constraints and objective are allowed to be nonlinear. Figure 2.1 shows a number of important models discussed in this book, and how they are related, where MINLP is the least restricted model, and LP is the most restricted. Edges in the figure represent additional restrictions on some aspect of the model. For instance, Linear Programming is a subset of Mixed Integer Linear Programming, in that all variables must be continuous. All Linear Programs are, in fact, Mixed Integer Linear Programs.

As computational environments have become more and more powerful, the types of models that can be efficiently processed have also grown to include complex models like MINLP, which have nonlinear and sometimes nonsmooth functions, coupled with discrete as well as continuous variables. Theoretical enhancements have extended the discipline beyond the convex domain to look at global solution of non-convex models. Even with tremendous strides, the field is still rapidly expanding and evolving to tackle difficult problems. The challenges of size (spatial/temporal/decision hierarchical) remain an active area of research: traditional approaches have proven inadequate, even with the largest supercomputers, due to the range of scales and prohibitively large number of variables. Furthermore, the nature of the model data remains problematic: how to deal with environments that are data sparse, data rich, or for which the underlying data is uncertain due to measurement errors, the lack of understanding of the model structure or random processes. Thus, rather than considering all possible optimization problems using the above features, we will specialize to a collection of problems that are useful to capture many of the aspects

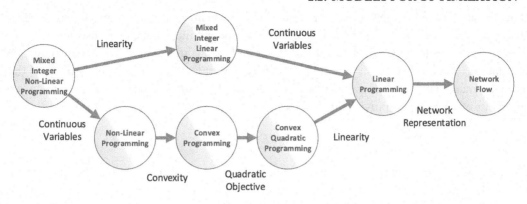

Figure 2.1: Optimization Models. Edges indicate additional restrictions to the model.

of a design optimization problem within the field of computer architecture. Table 2.1 outlines the optimization models we will cover in this chapter, defines their fundamental properties, and gives some example problems.

Optimization Model	Defined By	Examples
Linear Programming (LP)	Continuous Variables, Linear Constraints, Linear Objective	Resource Planning, Minimax Problems
Nonlinear Programming (NLP)	Continuous Variables, Nonlinear Constraints, Nonlinear Objective	Optimal Control, PDE Constrained Optimization
Convex Programming	Continuous Variables, Convex Domain, Convex Objective	Regression, Compressed Sensing, Second Order Cone Programming, Semidefinite Programing
Network Flow	Linear Programming, Network Representation & Constraints	Min Cost Network Flow, Shortest Path, Max Flow, Assignment Problems
Mixed Integer Linear Programming (MILP)	Continuous/Integer Variables, Linear Constraints, Linear Objective	Fixed Charge, Logic Programming, Ordering
Mixed Integer Nonlinear Programming (MINLP)	Continuous/Integer Variables, Nonlinear Constraints, Nonlinear Objective	Combines primitives of MILP, NLP

Table 2.1: Summary of optimization methods discussed in this chapter

2.2   MODELS FOR OPTIMIZATION

2.2.1   LINEAR PROGRAMMING

We begin by detailing the key ideas of optimization within the context of a linear program. Linear programs are the workhorse of optimization, much like solving linear systems is the workhorse of numerical analysis. Most optimization problems either use linear programming for subproblem solution, or are generalizations of the format that incorporate more complex constructs.

Linear programs involve variables, equations and inequalities. Variables are constrained by the equations and inequalities in the problem; these are written in terms of data (parameters) that a modeler must provide. Thus given parameters matrix A, vector b, Matrix E and vector g, the constraints of a linear program might look like:

$$Ax \geq b, Ex = g.$$

Theoretically, we can express the equation $Ex = g$ as two inequalities $Ex \geq g$ and $-Ex \geq -g$, and thus for simplicity of exposition we can simply consider the constraints to be $Ax \geq b$ without any loss of generality. The set $X = \{x \mid Ax \geq b\}$ is called the feasible region of the linear program: any point $x \in X$ satisfies the constraints of the design and is thus *feasible*. The feasible region defines (in a declarative fashion) constraints (or limits on the choice) of the whole collection of variables x. X is a polyhedral set: it is an intersection of half spaces; Figure 2.2 gives a pictorial view. We look for feasible points that optimize (i.e., maximize or minimize) some linear objective function $c^T x$ over X. To clarify, c^T is c transposed, which we use to perform the dot product

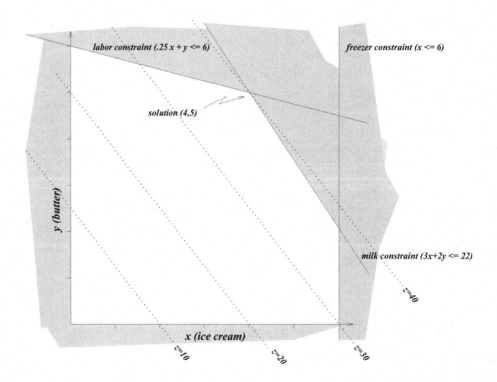

Figure 2.2: The feasible region of the Dairy Farmer Problem.

of $c^T x$, resulting in a scalar value. For exposition we shall assume a minimization, again without loss, since minimizing $c^T x$ is the same as maximizing $-c^T x$. Thus the general form of a linear program is (see Figure 2.3 for details):

$$\min_x c^T x \text{ s.t. } Ax \geq b. \tag{2.1}$$

Notice that X has extreme (corner or vertex) points: it can be shown that if X is non-empty, then either the problem is unbounded below (meaning the objective function can be driven to $-\infty$) or a solution exists that is an extreme point.

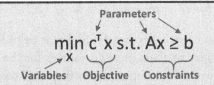

Above is the general form of linear program. Here x denotes the *variables*, or unknown quantities of the program. The *parameters*, c, A and b, are the inputs to the linear program, and define the specific instance. The constraints $Ax \geq b$ determine the feasible region, or domain of the problem. The *objective function*, $c^T x$ describes the characteristics of the best possible solution and the direction of optimization is signified using *max* or *min*.

Figure 2.3: General Form Linear Program.

To make the above description concrete, consider the Dairy Farmer Problem. A dairy farm makes ice cream and butter, and must decide how much of each to produce. Their cow produces 22 pints of milk per day, and it takes 3 pints to produce 1 pint of ice cream, and 2 pints of milk to produce 1 pint of butter. A pint of ice cream takes 15 min to make, while churning up a pint of butter takes an hour; there are 6 total hours of labor available. Finally the ice cream must be kept overnight in the freezer, and the total capacity is 6 pints. They can sell their product each morning for 4 dollars/pint for butter and 5 dollars/pint for ice cream. The objective is to maximize their total profit. This problem can be modeled aptly by a linear program, which is depicted in Figure 2.2. Here we have formulated resource constraints for total labor, milk production, and freezer space, limiting the total resources required in producing butter and ice cream to the maximum available. For instance, the "milk constraint" limits the production of butter and ice cream by enforcing that the total milk required (3 times the amount of ice cream, and 2 times the amount of butter) is less than the total milk available (22 pints). Also, dotted lines show the objective function, where each line represents choices which correspond to some overall profit. These are drawn according to the objective function (4 times the amount of butter, and 5 times the amount of ice cream). The solution is the feasible point which gives the highest profit.

Associated with any linear program is a dual linear program. The dual of (2.1) is

$$\max_{u} b^T u \text{ s.t. } A^T u = c, u \geq 0. \tag{2.2}$$

A critical component to the development of linear programming is the theory of duality.

Theorem 2.1 *Any linear program (2.1) is either infeasible, or its objective function can be driven to* $-\infty$ *(it is unbounded), or it has an optimal solution. When an optimal solution exists, then the dual problem (2.2) also has a solution, and their optimal objective function values are equal. Furthermore, any feasible point of the dual provides a lower bound on the optimal value of (2.1), and any feasible point of (2.1) provides an upper bound on the optimal value of (2.2).*

The theory of duality is an extremely powerful tool that helps to define algorithms for the solution of (2.1), and also provides mathematical guarantees for the existence of solutions. The dual variables u are important within economics where they are known as shadow prices due to the fact that the rate of change of the objective function of (2.1) with respect to changes in b_i can be shown to be u_i. Duality also indicates which constraints are important at solution points, and can be used as a practical (rigorous) mechanism to determine when to terminate an algorithm with a solution that is sufficiently close to being optimal.

We briefly comment on linear programming solvers here. There are two popular algorithm types for linear programs, namely the simplex method and interior point (or barrier) methods. The former searches for a solution by moving from one extreme point to the next, while the latter remains in the (relative) interior of the feasible region (by virtue of a barrier function that goes to infinity on the boundary of the feasible region) and traces a parametric path of solutions to an optimal feasible point. While the simplex method is exponential, linear programming is in complexity class $\mathcal{P}$; as variants of the interior point algorithm have polynomial complexity. In practice both methods are useful, since they have very different computational features that are relevant in varying situations.

2.2.2 CONVEX PROGRAMMING ★[1]

There has been a recent surge of interest in convex programming, which is a generalization of linear programming where the feasible region is a convex set, and the objective is any convex function. The relationship between convex and linear programming is shown in Figure 2.1.

A convex set C is a set that contains the line segments joining any two points in it, that is $C \subseteq \mathbb{R}^n$ (where $\mathbb{R}^n$ is the nth dimensional real domain) is convex if

$$x, y \in C \implies \{(1-\lambda)x + \lambda y \mid 0 \leq \lambda \leq 1\} \subset C.$$

Any polyhedral set (an intersection of half-spaces or linear constraints) is convex, and hence the feasible region of a linear program is convex. Convex functions can be defined by their epigraphs,

---

[1]★ indicates an optional section.

that is

$$\operatorname{epi} f = \{(x, \mu) \subset \mathbb{R}^{n+1} \mid f(x) \leq \mu\};$$

f is a convex function if its epigraph is a convex subset of $\mathbb{R}^{n+1}$. The convex programming problem is:

$$\min f(x) \text{ s.t. } x \in C, \tag{2.3}$$

where f is a convex function and C is a convex set. Therefore, linear programs are a special case of convex programs. Other cases include quadratic programming where

$$f(x) = \frac{1}{2} x^T Q x + c^T x$$

with Q being a positive semidefinite matrix ($x^T Q x \geq 0, \forall x$), and C being defined by linear (and convex quadratic) constraints. The factor $\frac{1}{2}$ is customary in the optimization literature since the matrix of second derivatives of f is used often and in this case will be equal to Q. This is purely a convenience.

Convex programming is a special case of a general nonlinear program, defined as:

$$\min f(x) \text{ s.t. } h_i(x) \leq 0, i = 1, \ldots, m, \tag{2.4}$$

where f and h_i are general nonlinear functions. Specifically, when f and h_i are convex functions, then (2.4) is a convex program (2.3), with $C = \{x \mid h_i(x) \leq 0, i = 1, \ldots, m\}$ being a convex set by virtue of each h_i being a convex function. Calculus plays an important role in the theory of nonlinear optimization. We note that this gives rise to optimality conditions (mechanisms to prove that a solution is optimal) but these are typically based on a local analysis. The point x is a local solution of (2.4) if there is some $\delta > 0$ such that

$$f(y) \geq f(x), \text{ whenever } y \in C \text{ and } \|y - x\| < \delta.$$

The key feature that makes convex programming attractive is the fact that any local solution of a convex program is in fact a global solution (where δ is infinite).

Much of the theory and algorithmic development of linear programming can be extended to the convex programming setting. In particular, the theory of duality extends in a natural way to this setting, albeit with some additional assumptions often termed constraint qualifications.

Minimax problems

We consider the solution of a modification of linear programming in which the linear objective function is replaced by a convex piecewise-linear function. Such a function can be represented as the pointwise maximum of a set of linear functions, that we can reduce to a linear programming problem. Figure 2.4 shows an example convex piecewise-linear function in blue, taken as the pointwise maximum of the linear functions shown in black.

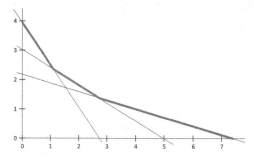

Figure 2.4: A convex piecewise-linear function.

Consider the function f defined as follows:

$$f(x) := \max_{i=1,\ldots,m} \left((c^i)^T x + d_i \right), \tag{2.5}$$

where $c^i \in \mathbb{R}^n$ and $d_i \in \mathbb{R}, i = 1, \ldots, m$. This function is convex. For f defined in (2.5), consider the problem

$$\min_x \ f(x), \ Ax \geq b. \tag{2.6}$$

By introducing an artificial variable μ, we can reformulate (2.6) as the following linear program:

$$
\begin{aligned}
\min_{(x,\mu)} \quad & \mu \\
\text{s.t.} \quad & \mu \geq (c^i)^T x + d_i, \quad \forall i = 1, \ldots, m; \\
& Ax \geq b.
\end{aligned}
$$

Note that the constraints themselves do not guarantee that μ equals $f(x)$; they ensure only that μ is *greater than* or equal to $f(x)$. However, the fact that we are *minimizing* ensures that μ takes on the smallest value consistent with these constraints, so at the optimum it is indeed equal to $f(x)$. The examples in Section 1.3.4 use this construction.

Any (continuous) convex function can be approximated to any degree of accuracy by a piecewise-linear convex function [96]. Thus, effectively a (separable) convex program can be approximated by a sequence of linear programs to any degree of accuracy using the technique outlined above. Note also that problems involving both $\|x\|_1$ (sum of absolute values of x) and $\|x\|_\infty$ (maximum element in x) can be expressed as linear programs using the formulation above.

2.2.3   NETWORK FLOW PROBLEMS ★

Network flow problems are a special case of the linear programming problem that can be solved more efficiently. We describe it here to give a flavor of the types of specialization of optimization that can occur, how that can be exploited computationally, and to show that solution properties of these restricted set of problems can be much better than the general form linear program.

Interestingly, the study of network flow models predates the development of linear programming. Seminal papers on transportation problems are [92, 101, 108], while the paper [153] provides even earlier references. The books by Dantzig [45] and Ford and Fulkerson [68] present thorough discussions of early contributions to this area, while [7] provides a more modern summary of the field.

Networks consist of a set of nodes $\mathcal{N}$ and arcs $\mathcal{A}$, where the arc (i, j) connects an origin node i to a destination node j; the set of arcs $\mathcal{A}$ is thus a subset of the Cartesian product $\mathcal{N} \times \mathcal{N}$. Schematically, nodes are normally represented by labeled circles; see Figure 2.5.

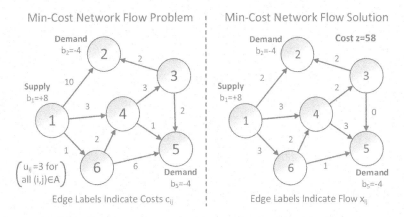

Figure 2.5: Example min-cost network flow problem and solution.

Associated with each node i is a *divergence* b_i, which represents the amount of product produced or consumed at node i. When $b_i > 0$, node i is a *supply node*, while if $b_i < 0$, it is a *demand node*. The variables x_{ij} in the problem represent the amount of commodity moved along the arc (i, j). Associated with each arc (i, j) are a lower bound l_{ij} and an upper bound u_{ij} of the amount of the commodity that can be moved along that arc. The unit cost of moving one unit of flow along arc (i, j) is c_{ij}. Typically, all the data objects c_{ij}, l_{ij}, u_{ij}, and b_i are assumed to be integral. The problem is to minimize the total cost of moving the commodity from the supply nodes to the demand nodes. Figure 2.5 shows an example problem with one supply node and two demand nodes.

Using the above notation, we can formulate the minimum-cost network flow problem as follows.

$$\min_{x} \quad z = \sum_{(i,j)\in\mathcal{A}} c_{ij} x_{ij}$$

$$\text{s.t.} \quad \sum_{j:(i,j)\in\mathcal{A}} x_{ij} - \sum_{k:(k,i)\in\mathcal{A}} x_{ki} = b_i, \quad \text{for all nodes } i \in \mathcal{N},$$

$$0 \le l_{ij} \le x_{ij} \le u_{ij}, \quad \text{for all arcs } (i, j) \in \mathcal{A}.$$

The first constraint states that the net flow through each node should match its divergence. The first summation represents the total flow *out of* node i, summed over all the arcs that have node

i as their origin. The second summation represents total flow *into* node i, summed over all the arcs having node i as their destination. The difference between inflow and outflow is constrained to be the divergence b_i.

The problem can be written in matrix form as the following linear program:

$$\begin{aligned} \min \quad & c^T x \\ \text{s.t.} \quad & \mathcal{I}x = b, \quad 0 \le l \le x \le u. \end{aligned} \tag{2.7}$$

Here, the *node-arc incidence matrix* $\mathcal{I}$ is an $|\mathcal{N}| \times |\mathcal{A}|$ matrix, the rows being indexed by nodes and the columns being indexed by arcs. Every column of $\mathcal{I}$ corresponds to an arc $(i, j) \in \mathcal{A}$ and contains two nonzero entries: a $+1$ in row i and a -1 in row j. For the example network of Figure 2.5, the matrix $\mathcal{I}$ is given by

$$\mathcal{I} = \begin{bmatrix} 1 & 1 & 1 & & & & & & \\ -1 & & & & -1 & & & & \\ & & & 1 & 1 & -1 & & & \\ & -1 & & & & 1 & 1 & -1 & \\ & & & -1 & & & -1 & & -1 \\ & & -1 & & & & & 1 & 1 \end{bmatrix},$$

where we have taken the arcs in the order

$$\{(1,2), (1,4), (1,6), (3,2), (3,5), (4,3), (4,5), (6,4), (6,5)\}.$$

Network flow problems of this nature are prevalent in the study of communication networks, on-chip and off-chip networks, and multi-commodity versions of this problem are used for routing of messages, for example. Other applications include vehicle fleet planning, building evacuation planning, karotyping of chromosomes, network interdiction, etc. In many examples, the problem doesn't quite fit into the formulation described above. In several cases, there are network transformations, described for example in [7], that massage the problem into the standard formulation.

For efficient implementations, it is crucial not to store or factor the complete matrix $\mathcal{I}$, but rather to use special schemes that exploit the special structure of this matrix. A node-arc incidence matrix is an example of a totally unimodular matrix, that is, a matrix for which the determinant of every square submatrix is equal to 0, 1, or -1. When A is totally unimodular and $\tilde{b}$, b, $\tilde{d}$ and d are integer vectors, it can be shown that if the set $\left\{ x \in \mathbb{R}^n \mid \tilde{b} \le Ax \le b, \tilde{d} \le x \le d \right\}$ is not empty, then all its extreme points are integer vectors (see [139] for further details). Since the simplex method moves from one extreme point to another, this result guarantees that the network simplex method will produce solutions x that are integer vectors whenever the data of the problem is integral.

We use this observation next to solve a number of problem types that occur frequently in large numbers of applications. While there are often specialized algorithms that can solve each of the following problems more efficiently than the way we outline, the formulations below are critical as component building blocks for the more general mixed integer linear programs (MILP) that we consider later. That is, if we use formulations of the type outlined in the remainder of this subsection within a more general MILP, then the relaxations that typical commercial grade solvers use are likely to perform significantly better than if we use a different (non-network) formulation.

Shortest Path Problem

The problem of finding the *shortest path* between two given nodes in the network can be formulated as a minimum-cost network flow problem by setting the costs on the arcs to be the distances between the nodes that the arc connects. If we wish to know the shortest path between nodes s and t, we set the divergences as follows: (we use \ as the set subtraction operator)

$$b_s = 1, \quad b_t = -1, \quad b_i = 0 \quad \text{for all } i \in \mathcal{N} \setminus \{s, t\}.$$

The lower bounds l_{ij} on the flows should be set to zero, while the upper bounds u_{ij} should be 1.

If we wish to know the shortest path from s to all the other nodes $i \in \mathcal{N}$, we define the network flow problem in the same way except that the divergences are different:

$$b_s = |\mathcal{N}| - 1, \quad b_i = -1 \quad \text{for all } i \in \mathcal{N} \setminus \{s\}.$$

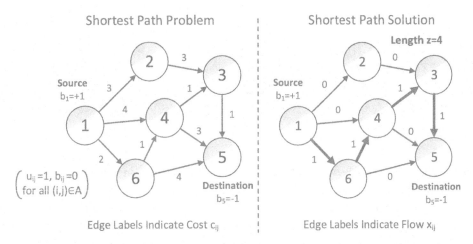

Figure 2.6: An example shortest-path problem and solution.

Figure 2.6 shows an example shortest-path problem and solution forlmulated as above. Having obtained the solution, we can recognize the shortest path from node s to a given node i by starting at i and backtracking along edges with positive flow.

Shortest path problems arise in many applications in the telecommunications, on-chip networks, and transportation industries whenever a message or a vehicle needs to be moved between two locations as quickly as possible, and as subproblems in other application such as project management and DNA sequencing. Specialized algorithms for the solution of this problem exist; the seminal reference is [50].

Max-Flow Problem

Given a network and two special nodes s and t, the *max-flow* problem is to determine the maximum amount of flow that can be sent from s to t. Of course, for the problem to be meaningful, some of the arcs in the network must have finite capacities u_{ij}.

This problem can be formulated as a minimum-cost network flow problem by adding an arc (t, s) to the network with infinite capacity, zero lower bound and a cost of -1. The divergences b_i at all the nodes are set to zero; the costs c_{ij} and lower bounds l_{ij} on all the original arcs are also set to zero. The added arc ensures that all the flow that is pushed from s to t (generally along multiple routes) is returned to s again and generates a profit (negative cost) corresponding to the flow on this arc. See Figure 2.7 below for an example.

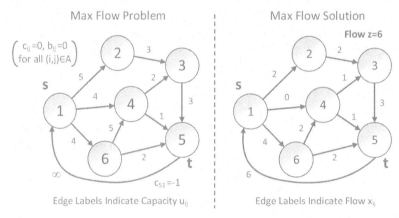

Figure 2.7: Example max-flow problem and solution.

We can also define max-flow problems with multiple sources, in which we wish to find the maximum amount of flow originating at any of the given sources that can be sent to the specified destination. To formulate as a minimum-cost network flow problem, we add a "super-source" as shown in Figure 2.8 and define arcs from the super-source to the original sources with infinite capacities and zero costs. We also add the arc from the sink to the super-source in the manner described above. Max-flow problems with multiple sinks (or multiple sources *and* multiple sinks) can be formulated similarly, see Figure 2.8.

Max-flow problem also occur in many application settings, including political redistricting, scheduling of jobs on parallel machines, assigning different modules of a program to minimize collective costs of computation and communication, and tanker scheduling.

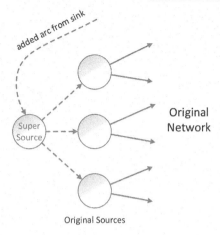

Figure 2.8: Replacing multiple sources by a super-source.

Assignment Problem

In the assignment problem we have two sets of nodes $\mathcal{N}_1$ and $\mathcal{N}_2$ of equal size. Given a specified cost c_{ij} for pairing a node $i \in \mathcal{N}_1$ with a node $j \in \mathcal{N}_2$, we wish to pair off each node in $\mathcal{N}_1$ with a partner in $\mathcal{N}_2$ (making a one-one correspondence between the two sets) so as to minimize the total cost of pairing. We can formulate this problem as a minimum-cost flow problem by defining the divergences as follows:

$$b_i = 1 \quad \text{for all } i \in \mathcal{N}_1, \quad b_i = -1 \quad \text{for all } i \in \mathcal{N}_2,$$

while the lower bounds and capacities are defined as follows:

$$l_{ij} = 0, \; u_{ij} = 1 \quad \text{for all } (i, j) \in \mathcal{A} \subset \mathcal{N}_1 \times \mathcal{N}_2.$$

Intuitively, the positive divergence for nodes in $\mathcal{N}_1$ forces flow along arcs to nodes in $\mathcal{N}_2$, which have negative divergence. The unit capacities on arcs make the assignment a one-to-one mapping. Figure 2.9 shows an example assignment problem and solution.

Assignment problems arise in a variety of problem contexts. Examples include personnel assignment, medical resident scheduling, locating objects in space, scheduling on parallel machines, and spatial architecture scheduling. A detailed treatment of assignment problems can be found in [35]. Network flow problems are used extensively in Internet traffic routing, and to model compiler flow graphs. Graph partitioning can also be approximated very effectively using network flow models, or when additional constraints are present, using a reformulation as a mixed integer program. We reiterate the point that MILP models built using underlying network structure are typically much more amenable to solution using the branch-and-bound or branch and cut procedures that are present in most commercial solvers.

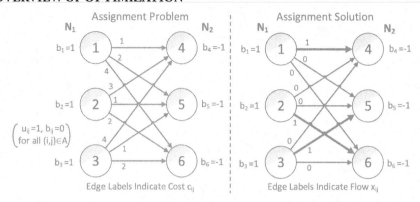

Figure 2.9: Example assignment problem and solution.

2.2.4 MIXED INTEGER LINEAR PROGRAMMING

The focus of this book is on linear programs in which a subset of the variables are constrained to take on only integer values. These models are called mixed-integer linear programs (MILP), and have the form:

$$\min_{x,y} \left(c^T x + d^T y\right) \text{ s.t. } Ax + Hy \geq b, x \in \mathbb{R}^n, y \in \mathbb{Z}^l \tag{2.8}$$

Notice that the variables are now separated into two sets, namely those that can take on continuous values (x), and those that can take only discrete (integer) values (y). The MILP acronym comes from the fact that there is a mix of continuous and discrete variables, and that all functional relationships are linear. Simple bound constraints $l \leq (x, y) \leq u$ can also be placed on the variables. In practice, it is important to specify good bounds (as tight as possible) on the integer variables since computational techniques rely heavily on such bounds.

Restricting some or all of the variables in a problem to take on only integer values can make a general linear program significantly harder to solve (more precisely, while linear programs are known to be polynomially solvable, mixed integer programs are NP-hard), so in a certain sense the result about the integrality of solutions to network linear programs is rather remarkable. Further information on this and other results can be found in the texts by Ahuja et al. [7], Nemhauser et al. [139], and Schrijver [152]. Unfortunately, the theory of duality also breaks down in this setting. While there are some duality results couched in the theory of submodular functions, this has not had anywhere near the impact that convex duality has on the field of convex optimization.

To give some intuition on what makes a MILP problem difficult to solve, we describe two fundamental properties of a MILP formulation, its convex hull and its linear relaxation. The *convex hull* is defined as the intersection of all convex sets containing the feasible set of the MILP (and hence is the smallest convex set containing the feasible set of the MILP). The *linear relaxation* of the MILP (2.8) simply relaxes the constraints $y \in \mathbb{Z}^l$ to be $y \in \mathbb{R}^l$, and hence is the linear program:

$$\min_{x,y} \left(c^T x + d^T y \right) \text{ s.t. } Ax + Hy \geq b, x \in \mathbb{R}^n, y \in \mathbb{R}^l$$

To explain, consider the following MILP, where x are integral.

$$\min_{(x_1,x_2) \in \mathbb{Z}^2} \quad x_1 + 3x_2$$

$$\begin{aligned}
\text{s.t.} \qquad x_1 - x_2 &\leq 2 \\
4x_1 - 11x_2 &\leq 20 \\
-2x_1 + 4x_2 &\leq 4 \\
x_1, x_2 &\geq 0
\end{aligned}$$

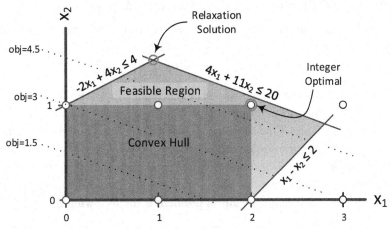

Figure 2.10: The convex hull of the feasible set of a MILP.

Figure 2.10 shows both the convex hull and feasible region for this problem. If the relaxed, linear problem is solved, then the optimal value obtained will not be integral. This solution is shown in the figure as the "relaxation solution." Often, the relaxation solution is rounded to the nearest integer. In some cases, this may be optimal, in other cases it may not be optimal or even be feasible for the original problem. In the given example, the rounded relaxation solution is $(1, 1)$, while the optimal solution is $(2, 1)$. If the convex hull of the feasible region of the MILP is identical to the feasible region of the linear relaxation, the relaxation is termed *tight*. In this case, solving the relaxed problem as a linear program would be sufficient, and computationally much less complex. Intuitively speaking, the closer or "tighter" the relaxed feasible region is to the convex hull, the easier it is to use the relaxed linear program as a way of helping to find the optimal integer solution, and the easier it is in general to solve the MILP. Solution methods are discussed further in Section 2.4. Though we don't define the term formally, "tightness" is sometimes used to refer to the how close the relaxation feasible region is to the convex hull.

As mentioned in the introduction, the last two decades have seen enormous improvement in the size and complexity of models that can be solved in realistic time frames in the MILP setting, and there are commercial and open source solvers that can be easily deployed in application domains. We will describe a number of problem formats that can be naturally expressed as MILPs in Section 2.3.

Due to the commonly used solution techniques, MILP is good for models for which the linear programming relaxation is tight, or for which the number of discrete choices is relatively small. MILP tends not to be as effective for ordering problems, or for complex very large scale models. There are many techniques that can be employed to extend the applicability of MILP to a stochastic setting, or to model both non-convex and nonsmooth phenomena. We also mention some of these techniques in Section 2.3.

2.2.5   MIXED INTEGER NONLINEAR PROGRAMS ★

Nonlinear models often arise in engineering. In many cases, an engineer has a view of her system as a collection of equations

$$F(x; p) = 0,$$

where p represents input values to the model, and x represents the state of the system; "scenarios" or "simulations" are performed to determine which of these input values leads to acceptable or good values for the state variables. One simple example of a nonlinear function would be modeling power as being proportional to the square of the voltage.

A concrete example of a nonlinear optimization problem is that of parameter fitting. In this setting we have a number of observations y_i of x and we wish to find the best values of p that explain all these observations. The optimization problem is the following instance of (2.4):

$$\min_{x,p} \sum_{i=1}^{m} \|y_i - x\|^2 \text{ s.t. } F(x, p) = 0.$$

There is a huge literature in statistics and optimization dealing with such regression or inverse problems (using observations of states to infer parameter values). In the data poor setting (m small) the problem is underdetermined (i.e., many x's achieve the minimum), and so modelers aim to impose additional structure in the formulation to compensate for the lack of data samples. Typically, this is an effort to allow for prediction and makes a trade-off between accuracy and a simple model structure (that can have more predictive power—Occam's razor). Compressed sensing and sparse optimization are burgeoning fields of exploration that have many examples of models of the form:

$$\min_{z} (E(z) + \alpha S(z)) \text{ s.t. } z \in X,$$

where X is the constraint set, E measures "error," and S penalizes bad structure. In the above example, we could set $z = (x, p)$, $X = \{(x, p) \mid F(x, p) = 0\}$, $E(z) = \sum_{i=1}^{m} \|y_i - x\|_2^2$ and

$S(z) = \gamma \|p\|_1$ resulting in the sparse optimization version

$$\min_{x,p} \sum_{i=1}^{m} \|y_i - x\|_2^2 + \gamma \|p\|_1 \text{ s.t. } F(x, p) = 0.$$

It should be noted that across disciplines, there are distinct terminology issues. For example, active learning is another term for optimal experimental design, and reinforcement learning is sometimes called approximate dynamic programming. However, it is widely acknowledged that incorporating domain knowledge into models (i.e., specifying F well) is critically important. Many opportunities remain to exploit theory and structure to generate much more effective algorithms, generalizability, and to understand learning behavior.

There is a large amount of recent work on the generalization of nonlinear programming problems (2.4) to include integer variables (a subset of the x variables are constrained to have integer values), so called mixed integer nonlinear programs (MINLP). At the present time these codes are less robust and unable to process models of the same size as MILP solvers. A key difficulty is that the "relaxed" problems remain difficult to solve to global optimality. The most widely used MINLP solvers currently appear to be BARON [162, 163] and LindoGlobal [118, 151], but BONMIN [29] and DICOPT [52] are also widely used but assume convexity of the underlying nonlinear program in order to guarantee global solutions.

2.3   MODELING PROBLEMS AS MILP

The expressive power of MILP means that it can capture a number of interesting behaviors and problem domains. In this section, we discuss a number of strategies for expressing "natural" design constraints to the MILP solver. Table 2.2 summarizes some general phenomena and the basic principle used to model them, and we describe the modeling in detail next. Note that the modeling of logical constraints is heavily employed in later chapters.

Modeling Problem	How To Formulate?
Logical Constraints	Use well-known transformations and binary variables
Ordering	Rank or explicit order formulation, or network flow
Piecewise-Linear Functions	Minimax (if convex objective), modeling line segments directly, or SOS2 variables
Mixed Integer Nonlinear Programs	Model nonlinearities with piecewise-linear functions, or convex envelopes
Mixed Integer Bilinear Programs	Use binary expansion formulation by [86]

Table 2.2: Summary of various modeling phenomena with MILP

2.3.1   LOGIC AND BINARY VARIABLES

We first give a simple example of how a binary variable δ can be used to impose a logical constraint. The simple case is often termed a "fixed charge" problem in the optimization literature. The idea is that if we use a machine, then we must start it up! Mathematically, we do this with the following logical implication:

$$x > 0 \rightarrow \delta = 1.$$

Here x represents the amount of production (for example) from a particular machine, so $x > 0$ captures the fact that we have produced something on the machine. The binary variable δ represents the "starting" of the machine, that is $\delta = 1$ precisely when the machine has been used. The logical constraint captures the fact that if we have produced something on the machine, then we must have started it up. Note in addition that if there is a startup cost c, then this can be modeled as an additional term in the objective function for example using the variable δ as $c\delta$. The above implication has two issues. First, it involves a strict inequality, and the linear and convex formulations outlined earlier in this chapter are all written in terms of equations and (non-strict) inequalities (essentially so the feasible region is a closed set). Secondly, logical implications are not immediately expressible as an equation or inequality, strict or otherwise. One way around this introduces a parameter M, an upper bound on the variable x, and then replaces the logical implication by

$$x \leq M\delta.$$

If $x > 0$, this inequality forces δ to take on the value 1. Any model that includes the logical implication, simply replaces it with the latter inequality. The reverse implication

$$\delta = 1 \rightarrow x > 0$$

requires an additional fix, that is an $\epsilon > 0$ with the property that we replace $x > 0$ with $x \geq \epsilon$ for suitably chosen (small) ϵ. In this case, provided $x \geq 0$, the implication $\delta = 1 \rightarrow x \geq \epsilon$ is equivalent to

$$x \geq \epsilon\delta.$$

The following theorem captures concisely some possible generalizations of this idea. Each of the statements first has an upper or lower bound expression that must be satisfied at any feasible solution (i.e., m or M must be chosen so that the given expression is automatically satisfied). The remainder of the statement then gives the logical expression on the left, followed by the constraint that must be added to any MILP (when g is linear) to model that logical expression. Note that the case (1b) is a generalization of the fixed charge case outlined above ($g(x) = x$).

Theorem 2.2   *Let $x \in \mathbb{R}^n$ and $g: \mathbb{R}^n \rightarrow \mathbb{R}$.*

1. *Suppose for all x, $g(x) \leq M$. Then,*

    (a) $(\delta = 1 \rightarrow g(x) \leq 0) \iff g(x) \leq M(1 - \delta)$

(b) $(g(x) > 0 \to \delta = 1) \iff g(x) \le M\delta$

(c) Suppose $\epsilon > 0$ is small enough so that $g(x) < 0 \iff g(x) \le -\epsilon$. Then
$(g(x) \ge 0 \to \delta = 1) \iff g(x) \le (M + \epsilon)\delta - \epsilon$

2. Suppose for all x, $g(x) \ge m$. Then,

(a) $(\delta = 1 \to g(x) \ge 0) \iff g(x) \ge m(1 - \delta)$

(b) $(g(x) < 0 \to \delta = 1) \iff g(x) \ge m\delta$

(c) Suppose $\epsilon > 0$ is small enough so that $g(x) > 0 \iff g(x) \ge \epsilon$. Then
$(g(x) \le 0 \to \delta = 1) \iff g(x) \ge (m - \epsilon)\delta + \epsilon$

The parts of this theorem can be combined to model implications of the form $f(x) > 0 \to g(x) \le 0$ using

$$(f(x) > 0 \to \delta = 1) \text{ and } (\delta = 1 \to g(x) \le 0).$$

Similarly $g(x) = 0 \leftrightarrow \delta = 1$ can be modeled when g is linear within a MILP using:

$$g(x) \le (M + \epsilon)\delta_1 - \epsilon$$
$$g(x) \ge (m - \epsilon)\delta_2 + \epsilon$$
$$\delta_1 + \delta_2 \le 1 + \delta$$

As we will see in the case studies, constraints of this form are extremely useful in expressing natural phenomena and system requirements.

Solvers often introduce additional variable types to capture some of these notions; for example, semicontinuous variables are variables that are either 0 or lie between given positive bounds. Another type of variable that is common among many solvers is an SOS1 variable (specially ordered set of type 1). This is a collection of variables defined over an (ordered) set K, at most one of which can take a strictly positive value, all others remaining 0. Branching strategies (see Section 2.4.1) can exploit the ordering in K, for example when it indicates a choice between small, medium, large, and super-sized items. This can, in some cases, lead to improved solution times if such semantics are required.

2.3.2 CONSTRAINT LOGIC PROGRAMMING

We let binary variables δ_i represent propositions P_i via the following construction:

$$\delta_i = \begin{cases} 1 & \text{if proposition } P_i \text{ is true} \\ 0 & \text{if proposition } P_i \text{ is false} \end{cases}$$

In this case, δ_i is often termed an indicator variable since it indicates whether the proposition is true or false. It is also used in statistics to indicate whether a variable takes on a particular value (or set of values) or not.

We will use standard notation from boolean algebra to denote connectives between propositions. Thus,

$\vee$	means 'or'
$\wedge$	means 'and'
$\neg$	means 'not'
$\rightarrow$	means 'implies'
$\leftrightarrow$	means 'if and only if'
$\veebar$	means 'exclusive or'

Other connectives such as "nor" or "nand" are also used in the literature.

The proposition P_i could stand for "we will use register i", and Q could represent "perform compiler level 3 optimization", so that $Q \rightarrow P_i$ encodes a logical constraints that if we use a level 3 optimizing compiler then we must use register i. A key part of modeling that we demonstrate in examples later in this book is determining what variables capture the underlying logic of our problem, and finding the connectives that they must satisfy in order for the design to capture the underlying required properties.

Table 2.3 details standard ways to equivalently express propositional logic in terms of constraints on the corresponding indicator variables in a MILP (see [134] for example). The examples shown in the table are useful in building models since they construct a tight approximation of the logic typically, even when the solution algorithm used to solve the MILP relaxes some of the variables to be continuous (i.e., in [0, 1] instead on being in {0, 1}).

Other operators like "before", "last" or "notequal", "allDifferent" are often allowed in constraint logic programming (CLP) languages; there is a growing literature on how to reformulate some of these within a MILP code and lots of specialized codes that treat these constraints explicitly. Merging these two techniques (MILP and CLP) is an active area of research. The techniques used in CLP are essentially clever ways to do complete enumeration very efficiently and quickly.

2.3.3 ORDERING

Ordering, in this context, refers to the arrangement of events in a certain domain, subject to some constraints. An example from compilers would be the ordering of instructions inside a software pipelined loop for maximum throughput.

The constraint logic programming constructs are particularly useful in ordering problems since they can easily encode the notions of an ordering. We outline here two of the main ideas applicable when a pure MILP approach is used. These are only practical on medium sized problems at this time, although much research is currently underway to improve the size of problems that are practically tractable.

One formulation of ordering uses binary variables rank with the definition:

$$\text{rank}_{ik} = 1 \text{ if item } i \text{ has position } k.$$

Statement	Constraint
$\neg P_1$	$\delta_1 = 0$
$P_1 \vee P_2$	$\delta_1 + \delta_2 \geq 1$
$P_1 \veebar P_2$	$\delta_1 + \delta_2 = 1$
$P_1 \wedge P_2$	$\delta_1 = 1, \delta_2 = 1$
$\neg(P_1 \vee P_2)$	$\delta_1 = 0, \delta_2 = 0$
$P_1 \rightarrow P_2$	$\delta_1 \leq \delta_2$ [equivalent to: $(\neg P_1) \vee P_2$]
$P_1 \rightarrow (\neg P_2)$	$\delta_1 + \delta_2 \leq 1$ [equivalent to: $\neg(P_1 \wedge P_2)$]
$P_1 \leftrightarrow P_2$	$\delta_1 = \delta_2$
$P_1 \rightarrow (P_2 \wedge P_3)$	$\delta_1 \leq \delta_2, \delta_1 \leq \delta_3$
$P_1 \rightarrow (P_2 \vee P_3)$	$\delta_1 \leq \delta_2 + \delta_3$
$(P_1 \wedge P_2) \rightarrow P_3$	$\delta_1 + \delta_2 \leq 1 + \delta_3$
$(P_1 \vee P_2) \rightarrow P_3$	$\delta_1 \leq \delta_3, \delta_2 \leq \delta_3$
$P_1 \wedge (P_2 \vee P_3)$	$\delta_1 = 1, \delta_2 + \delta_3 \geq 1$
$P_1 \vee (P_2 \wedge P_3)$	$\delta_1 + \delta_2 \geq 1, \delta_1 + \delta_3 \geq 1$

More general forms of some of the above are also stated below:

$P_1 \vee P_2 \vee \cdots P_n$	$\sum_{i=1}^{n} \delta_i \geq 1$
$(P_1 \wedge \cdots P_k) \rightarrow (P_{k+1} \vee P_n)$	$\sum_{i=1}^{k}(1 - \delta_i) + \sum_{i=k+1}^{n} \delta_i \geq 1$
at least k out of n are true	$\sum_{i=1}^{n} \delta_i \geq k$
exactly k out of n are true	$\sum_{i=1}^{n} \delta_i = k$
at most k out of n are true	$\sum_{i=1}^{n} \delta_i \leq k$
$P_n \leftrightarrow (P_1 \vee \cdots \vee P_k)$	$\sum_{i=1}^{k} \delta_i \geq \delta_n, \delta_n \geq \delta_j, j = 1, \ldots, k$
$P_n \leftrightarrow (P_1 \wedge \cdots \wedge P_k)$	$\delta_n + k \geq 1 + \sum_{i=1}^{k} \delta_i, \delta_j \geq \delta_n, j = 1, \ldots, k$

Table 2.3: Propositional logic as MILP constraints

The model then includes assignment type constraints that indicate each position k contains one item, and each item i has exactly one rank:

$$\sum_i \text{rank}_{ik} = 1, \forall k \text{ and } \sum_k \text{rank}_{ik} = 1, \forall i.$$

It is then fairly straightforward to generate expressions for entities such as the start time of the item in position k, for example, and thereby generate expressions for waiting time and other objectives.

A second type of formulation involves ordering. The model implements the condition that either item j finishes before item i starts or the converse. Such either-or constraints are termed disjunctions. To represent the start and stop times of a particular task, we use the variables $start_i$ and end_i. We introduce additional "violation" variables that measure how much the pair (i, j) violates the condition that j finishes before i starts (and the converse):

$$end_i \leq start_j + \text{violation}_{ij} \text{ and } end_j \leq start_i + \text{violation}_{ji}$$

We then add a condition that only one of these two variables—violation_{ij} and violation_{ji}—can be positive, that is we force each such pair of variables to be in an SOS1 set of size 2 of the form $\{\text{violation}_{ij}, \text{violation}_{ji}\}$. Some solvers (e.g. CPLEX) implement the notion of indicator constraints which provide an alternative way to formulate disjunctions.

Network flow problems can also be used for ordering problems, often with great success. In many cases, the graph formulation allows the ordering constraints to be modeled in a more computationally effective manner. Examples tend to be very domain specific; they often enumerate many different "ordered paths" and use a flow formulation to select among those paths.

2.3.4   PIECEWISE-LINEAR MODELS ★

Recall the general nonlinear program (2.4):

$$\min f(x) \text{ s.t. } h_i(x) \leq 0, i = 1, \ldots, m.$$

Models of this nature appear throughout the literature and there are a number of excellent solvers that are effective on even large scale instances of them. Most of these solvers find local solutions however, and the extension to non-convex models is known to be NP-hard even when all the functions are univariate (depending on only a single variable)[104].

Piecewise-linear functions are extensively used to approximate the original functions in the non-convex setting. There are a large number of applications of this idea, along with specialized algorithms and reformulations as MILP. We refer the reader to [172] for more extensive references.

We explain this approximation first in the case of a function of a single variable x. We restrict attention to the non-convex case since the convex case can be solved using the approach outlined in Section 2.2.2 for minimax problems.

The piecewise-linear function is described by a collection of segments $\mathcal{S}$. In the case where the domain of the function is an unbounded set, or the function is not continuous, the segment approach has proven effective. Each segment i has an (x_i, f_i) coordinate point, a (potentially infinite) length l_i, and a slope g_i, the rate of increase or decrease of the function from (x_i, f_i). The sign of the l_i determines if the segment expands to the left (negative length) or the right

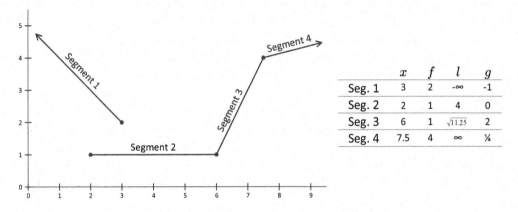

	x	f	l	g
Seg. 1	3	2	$-\infty$	-1
Seg. 2	2	1	4	0
Seg. 3	6	1	$\sqrt{11.25}$	2
Seg. 4	7.5	4	∞	¼

Figure 2.11: An example of a (multi-valued) piecewise-linear function described by segments.

(positive length) of the (x_i, f_i) point. These segment definitions allow more than pure piecewise-linear functions. Segments can overlap, meaning we can have multi-valued functions, and there can be holes in the x coordinate space. There is also no order requirement of the segment x_i coordinates.

Each segment has two variables associated with it. The first is a binary variable b_i that chooses the segment to be used. In order that we have a single value for the function at x, only one segment can be active, which is modeled using:

$$\sum_{i \in \mathcal{S}} b_i = 1.$$

The other segment variable is a nonnegative variable λ_i whose upper bound is the absolute value of the length of the segment: $\lambda_i \le |l_i|$. This variable measures how far we move into segment i from the starting point (x_i, f_i). A particular choice of the vectors b and λ formed from these components determines a point of evaluation $x \in \mathbb{R}$ and the value of the approximation f at x by the following formulae ($sgn(l_i)$ denotes the "sign" of the parameter l_i):

$$x = \sum_{i \in \mathcal{S}} (x_i b_i + sgn(l_i) \lambda_i), \quad f = \sum_{i \in \mathcal{S}} (f_i b_i + sgn(l_i) g_i \lambda_i).$$

For each segment that has finite length $|l_i| < \infty$, we enforce the constraint that λ_i can only be positive if $b_i = 1$ using the M constraint:

$$\lambda_i \le |l_i| b_i.$$

If the piecewise-linear function contains segments of infinite length, this constraint does not work. Instead, for these segments, we form a SOS1 set containing the variables λ_i and $1 - b_i$, that is at most one of these two variables is positive. This has the same effect as the M constraint, but is independent of the length of the segment and hence also works with infinite length.

Figure 2.11 gives an example piecewise-linear multi-valued function, and shows the associated parameters which define the lines. Note that the function values between $x = 2$ and $x = 3$ are not determined (the function is multi-valued there). It can take on either the value 1, or the value $5 - x$. The optimization procedure will choose the preferable value.

In the non-convex setting there are other popular MILP formulations for piecewise-linear functions namely incremental cost [103, 125] and convex combination [44]. Historically, [18] suggested a formulation for piecewise-linear functions similar to convex combination, except that no binary variables are included in the model and the nonlinearities are enforced algorithmically, directly in the branch-and-bound algorithm, by branching on sets of variables, which they called special ordered sets of type 2 (SOS2). It is also possible to formulate piecewise-linear functions similar to incremental cost but without binary variables and enforcing the nonlinearities directly in the branch-and-bound algorithm. Two advantages of eliminating binary variables are the substantial reduction in the size of the model and the use of the problem structure [63]. Since these variables may reduce the preprocessing opportunities, it may be that the reduction in size of the problem does not lead to overall solution speedup.

We now outline some of the ways the single variable case can be extended to a more general setting.

Separable programming

A function f is called separable if it can be expressed as the sum of functions f_j of a single variable $x_j \in \mathbb{R}$:

$$f(x) = \sum_j f_j(x_j)$$

The nonlinear optimization problem (2.4) is separable if f and h_i are all separable functions.

If the problem is not separable, there are a number of tricks that can be used to substitute out non-separable terms and convert the model into a separable one (see [178]). For example, we can deal with terms like $x_i x_j$ by using the fact that

$$4x_i x_j = (x_i + x_j)^2 - (x_i - x_j)^2$$

or terms like

$$\prod_{i=1}^m x_i$$

(with $x_i > 0$) can be replaced with y where

$$\ln y = \sum_{i=1}^{m} \ln x_i.$$

Note that linear functions are separable, so functions like $f(\sum_j a_j x_j)$ can be reformulated in a separable manner using $f(y)$ where $y = \sum_j a_j x_j$.

For a specific example, consider the problem:

$$\min_{x_1, x_2} x_1 x_2 - \log(x_1 + 2x_2) \text{ s.t. } x_1^2 + x_2^2 \leq 1.$$

Using the product form reformulation and introducing additional constraints and variables the problem can be made separable as follows:

$$\min_{x_1, x_2, w, y, z} \frac{1}{4} y^2 - \frac{1}{4} z^2 - \log(w) \text{ s.t. } x_1^2 + x_2^2 \leq 1, y = x_1 + x_2, z = x_1 - x_2, w = x_1 + 2x_2.$$

However, in general this may lead to a large growth in the number of variables and constraints in the resulting model. Once a problem has been converted into separable form, the separable programming technique basically replaces all separable functions, in objectives and constraints, by piecewise-linear functions.

The epigraph of a piecewise-linear function is easily seen to be the union of polyhedra. It is possible to approximate a non-separable function by a general function of this form. The paper [172] gives an excellent treatment of how mixed integer models can approximate the general non-separable case, with pointers to which formulation is best in what setting.

2.3.5 MODELING MIXED INTEGER NONLINEAR PROGRAMS ★

A popular modeling approach for solving MINLP is to approximate the nonlinear functions by piecewise-linear approximations as outlined above and thus approximate the MINLP by a MILP. If the function is convex, then the minimax approach outlined in Section 2.2.2 is one method that can be used—in this setting there is no need to introduce additional discrete variables since the optimization itself guarantees the approximation is tight. Other techniques generate convex underestimators of the objective function based on ideas of factorable functions and cutting planes [127, 129].

Many of the nonlinearities that appear in integer programming are in the form of polynomial functions, and many of these involve terms no higher than second order. A standard approach to linearizing these polynomial functions is to expand each integer variable y using a collection of binary variables ($y = \sum_{i=1}^{k} 2^{i-1} x_i$, x_i binary) and then to introduce new binary variables to take the place of the cross product terms. Note that if $w = x_1 * x_2$, where $x_i \in \{0, 1\}$, then the inequalities $w \leq x_1$, $w \leq x_2$ and $w \geq x_1 + x_2 - 1$ can replace the product definition of w and ensure that w will assume the appropriate values. The following section outlines how this is performed in a particular setting. Improvements and specializations for specific structures are outlined in the seminal work [76].

Modeling Mixed Integer Bilinear Programs

In many application settings, the following extension of MILP arises:

$$\min_{x,y} \quad x^T Q_0 y + c_0^T x + d_0^T y$$
$$\text{s.t.} \quad Ax + Hy \geq b_0$$
$$\qquad x^T Q_t y + c_t^T x + d_t^T y \leq b_t, \quad t = 1, \ldots, m$$
$$\qquad 0 \leq x \leq a$$
$$\qquad 0 \leq y \leq u, y \in \mathbb{Z}^l$$

Crucially, both the continuous and the discrete variables are bounded. A very effective reformulation of this problem carries out a binary expansion of the variable y_j as $y_j = \sum_{i=1}^{k} 2^{i-1} z_i$ with z_i binary (and suitably chosen k) and then replaces the product terms $x_l z_j$ by a new variable w_{lj}, which is also expanded using additional variables v. It can be shown that $(x_l, y_j, w_{lj}, z_j, v_{lj}) \in \mathcal{B}_{lj}$ and that $\mathcal{B}_{lj}$ is a polyhedral set with some integer restrictions (on y_j and z_j), and that solving

$$\min_{x,y,w,z,v} \quad \sum_l \sum_j (Q_0)_{lj} w_{lj} + c_0^T x + d_0^T y$$
$$\text{s.t.} \quad Ax + Hy \geq b_0$$
$$\qquad \sum_l \sum_j (Q_t)_{lj} w_{lj} + c_t^T x + d_t^T y \leq b_t, t = 1, \ldots, m$$
$$\qquad (x_l, y_j, w_{lj}, z_j, v_{lj}) \in \mathcal{B}_{lj}$$

gives a solution to the original problem. This problem is a MILP. An excellent reference for this material and its extensions is [86]. In particular, the paper shows the above reformulation to be very effective under modern MILP solvers, and that new cuts for $\mathcal{B}$ can be derived to make the solution procedure even more efficient. There is no restriction on Q being positive semidefinite in this setting.

An implementation of many different reformulations for problems of this form is provided by the GloMIQO solver [131] that is available within the GAMS modeling system.

2.4   SOLUTION METHODS ★

There is vast literature on algorithms for solving optimization problems. That literature includes theoretical work devoted to issues of existence and uniqueness of solutions, stability analysis related to perturbations of the data defining the problems, speed of convergence of algorithms and complexity of the underlying problem classes. All of this theory is important and extremely relevant to optimization, but in the interest of space we omit it here. The interested reader is pointed to references such as [139, 149, 152].

In numerical analysis, the key building block for solving general systems of equations are methods for solving linear equations (such as LU factorization or iterative schemes such as Krylov subspace methods). The key building block for solving mixed integer linear programs is linear programming. Linear programs are well solved using commercial packages such as CPLEX, Gurobi and Xpress, as well as academic codes such as CLP, MINOS, and others. All of these utilize modern sparse linear algebra techniques, and the commercial packages incorporate sophisticated presolving techniques (such as probing, aggregation, forcing) that help reduce the size of the problem being solved and hence improve efficiency. Commercial codes include options to apply the primal or dual simplex method, automatically extract network structure and solve using the network simplex algorithm, or use a barrier method. For large scale settings, the barrier method is often more effective than the simplex method, and the advantages of a basic solution are also made available by a crossover method. Simplex methods are much better at restarting from a given starting point.

Solvers can be tuned and tweaked for improved performance using a variety of techniques, some of which are mentioned in Chapter 7. Critical to understanding what a solver is doing is the log file produced during the execution of the optimization. These are very specific to each solver, however, and the reader is referred to the online documentation [74] for further details.

2.4.1 BRANCH-AND-BOUND

The standard method for solving MILP is branch-and-bound, which uses the linear programming methodology outlined above to solve many subproblems. In this setting, (bounded) discrete variables are replaced by continuous variables, and the resulting (so called root-relaxation) problem is solved using a linear programming code. The optimal value of this relaxation provides a lower bound on the optimal value of the MILP. If at a solution, all the continuous replacement variables take on valid discrete values, then the original problem is solved. Otherwise, one of the integer variables x_i has the value $\bar{x}_i$ which is not integral and is branched on: that is, two (MILP) subproblems are generated, one of which adds the constraint that $x_i \leq \text{floor}(\bar{x}_i)$ and the other adds the constraint $x_i \geq \text{ceil}(\bar{x}_i)$. The variable x_i is called the branching variable. If we can compute the optimal solutions for both of these subproblems then the better of them will be the solution of the original MILP. In this way a search tree of nodes is constructed, each node consisting of an MILP that is the original MILP augmented with a collection of bound restrictions on variables. We can repeat this process: at any stage of the algorithm, we have a search tree, each of whose leaves are MILP subproblems (augmented with additional linear constraints) that are not solved. If we can solve or discard all of these leaves, then the original MILP is solved by one of the leaf solutions.

Consider the following example problem, a variant of the $0 - 1$ knapsack problem.

$$\begin{aligned}
\min_{x} \quad & -8x_1 - 6x_2 - 5x_3 - 4x_4 \\
\text{s.t.} \quad & 5x_1 + 4x_2 + 4x_3 + 2x_4 \leq 8 \\
& x_1, x_2, x_3, x_4 \in \{0, 1\}
\end{aligned}$$

The branch-and-bound procedure for this problem can be seen in Figure 2.12. We start by having no incumbent solution, which is defined as the current best solution which meets the integrality constraints. Therefore, the upper bound best solution is ∞. In Step 1, we solve the root relaxed linear program, and achieve a fractional solution ($x_2 = 1/4$). The objective for this problem, −13.5, serves as a lower bound for any solution attainable. To continue, we create two new problems by forcing x_2 to either 0 or 1. Both of these problems again have fractional solutions.

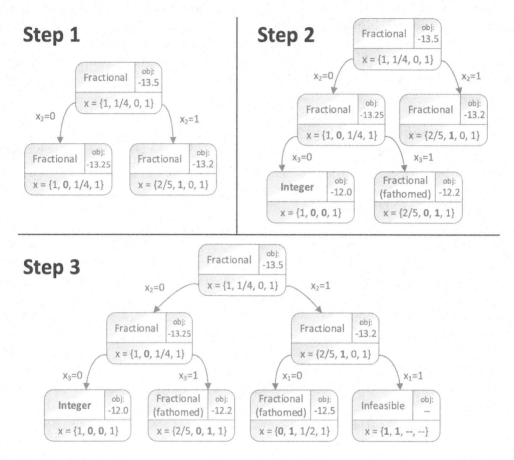

Figure 2.12: Branch-and-bound tree for previous MILP.

When multiple subproblems are available for branching, a node selection strategy is followed; one method is to choose the solution with smaller node relaxation objective function. Therefore, in Step 2, we create two new problems from the subproblem with objective −13.25, by fixing the value of x_3.

We now describe the process of fathoming (discarding nodes). Suppose an unsolved leaf MILP subproblem is selected for solution. We relax the integrality constraints on the subproblem and solve the node relaxation. If the node relaxation solution is feasible for the original MILP problem, then we update the incumbent solution to be the best among this solution and the existing incumbent and we can fathom this node if its solution is worse. In Step 2, with $x_3 = 0$ we achieve an integer solution with objective value −12. This solution becomes our incumbent.

The second option is that the node relaxation is solved (it cannot be unbounded since otherwise its ancestor node would not have been solvable), and we can fathom this node if its optimal value is larger than the incumbent value (since branching on this node will not beat the incumbent). For example, in Step 2, the subproblem with $x_3 = 1$ has objective value -12.2. However, since all of the variables and parameters in the objective are integral, a sophisticated solver will realize that this can also be fathomed. This is because the best objective value that this subproblem can give is -12 when rounded up to the nearest integer, and our incumbent integer solution has objective -12.

The third option is that the node relaxation is infeasible (we are after all adding constraints to it) so again that node is fathomed. (This occurs in our example in Step 3, when we consider the node generated by setting the variable $x_1 = 1$.) Otherwise, we branch on this node to create two new nodes.

Even though an integer solution has been found, we have not proven optimality until no subproblem has an objective value which is lower than the current incumbent. So in Step 3, we return to the previous leaf (where $x_2 = 1$), and continue by fixing the value of x_1. When x_1 is fixed to 0, the objective is -12.5, and can therefore be fathomed by the same argument as before. When x_1 is fixed to 1, the problem becomes infeasible as mentioned above. Since all subproblems are now either integral, fathomed, or infeasible, the procedure is complete, and we have found the optimal value.

In general, this process may take a very long time to complete. The remaining issue is to explain how we stop the process in practice: this is accomplished by determining tight upper and lower bounds on the optimal solution value of the MILP. Note that the value of the incumbent solution is an upper bound on the optimal objective of the MILP (we are minimizing after all). A lower bound can also be found: it is the minimum of the objective values of the node relaxations over all the leaf nodes of the search tree. This value can be updated as we proceed in the algorithm: the difference between the upper and lower bound is typically called the gap and we often curtail the algorithm when this gap gets sufficiently small. Thus, in our example, if we were only looking for a solution which was 15% within optimal, we could have stopped when we found the first integer solution. The ability to control the degree of optimality for a solution is one of the primary benefits of MILP, and mathematical optimization in general.

2.4.2 EXTENSIONS TO BASIC BRANCH-AND-BOUND

Modern MILP codes extend the basic branch-and-bound scheme in many different ways. One principal way is via the addition of cutting planes, or cuts. The core idea is that if we could describe the convex hull (the smallest convex set containing all) of the integer feasible points of the MILP, then solving a linear program over that convex hull would give the optimal solution of the MILP. Essentially, this approach adds constraints to make the formulation "tighter."

A cut is a hyperplane that separates the solution of the node relaxation problem from this convex hull. The best cuts are those that form facets of the convex hull. As an example, Figure 2.13

shows the same problem as Figure 2.10, with the addition of two cutting planes, at $x_1 \leq 2$ and $x_2 \leq 1$. Advanced codes try a variety of different choices of cuts in an attempt to find the best ones for the particular problem.

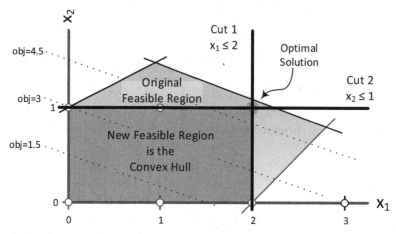

Figure 2.13: Adding cutting planes to a MILP to improve the formulation.

Another extension to branch-and-bound is to employ presolve techniques. Presolve is a collection of problem reductions that are carried out before starting the branch-and-bound process. These reductions aim to reduce the size of the problem and to tighten its formulation to conform more closely to the convex hull, and include procedures to eliminate redundant constraints, fix variables, scale or reduce coefficients, and improve bounds on variables. Heuristics are procedures that try to find good integer feasible points, and these are useful as incumbent solutions (and hence as upper bounds). Other options include mechanisms to choose the branching variable and the node to process next, detect symmetry between solutions, and to solve nodal relaxations in parallel.

Lagrangian relaxation is a technique for generating better lower bounds on the optimal value of the MILP. The seminal paper on this approach is [66]. It is a technique that can be applied to problems whose constraints can be divided into two sets, namely "good" constraints (with which the problem is easily solved) and "bad" constraints that make it very hard to solve. The main idea is to remove the bad constraints by putting them in the objective function and penalizing their violation with weights (called Lagrangian multipliers). The multipliers get updated in an iterative manner.

The overall approach can be described as follows. Consider the MILP:

$$z = \min_{x} c^T x \text{ s.t. } Ax \leq b, x \in Y$$

where the set Y involves a collection of integrality constraints along with linear restrictions. The constraints $Ax \leq b$ are often referred to as complicating (bad) constraints: if we relax them the problem

$$\min_x c^T x \text{ s.t. } x \in Y$$

is assumed to be relatively easy to solve (Y encodes the good constraints), even though it is a MILP. We exploit this fact by constructing the *Lagrangian Dual:*

$$\mathcal{L}(u) := \min_x \left(c^T x + u^T (Ax - b)\right) \text{ s.t. } x \in Y$$

(so that for fixed u we can evaluate $\mathcal{L}(u)$ easily), and note that for any $u \geq 0$, $\mathcal{L}(u)$ forms a lower bound on z (the optimal value of the MILP) since for any feasible x, $u^T (Ax - b) \leq 0$. Lagrangian relaxation methods aim to find strong lower bounds by solving

$$\max_{u \geq 0} \mathcal{L}(u).$$

Similar relaxation schemes can be used for convex optimization, including those that are based on semidefinite programming.

A specific example is the generalized assignment problem:

$$\min_x \sum_{i=1}^{m} \sum_{j=1}^{n} c_{ij} x_{ij} \text{ s.t. } \sum_{j=1}^{n} a_{ij} x_{ij} \leq b_i, i = 1, \ldots, m, x \in Y$$

where

$$Y = \left\{ x = (x_{ij}) \mid \sum_{i=1}^{n} x_{ij} = 1, \forall j, x_{ij} \in \{0, 1\}, \forall i, j \right\}.$$

In this case, the Lagrangian relaxation problem (for fixed $u \geq 0$) is

$$\min_{x \in Y} \sum_{i=1}^{m} \sum_{j=1}^{n} c_{ij} x_{ij} + \sum_{i=1}^{n} u_i (\sum_{j=1}^{n} a_{ij} x_{+ij} - b_i)$$

$$= \min_{x \in Y} \sum_{i=1}^{m} \sum_{j=1}^{n} (c_{ij} + u_i a_{ij}) x_{ij} - \sum_{i=1}^{n} u_i b_i$$

These subproblems are solved in time proportional to nm by determining $\min_i (c_{ij} + u_i a_{ij})$ for each j, setting the associated $x_{ij} = 1$ and all the remaining x_{ij} to zero. (Subgradient optimization approaches can then be used to update the vector u.)

2.4.3 COLUMN GENERATION

Column generation is a method for solving large scale linear programs, where the number of variables is much larger than the number of constraints. The core idea is to select a subset $J \subseteq \{1, \ldots, n\}$ of the variables and to solve the approximate (restricted master) problem:

$$\min_x \sum_{j \in J} c_j x_j \text{ s.t. } \sum_{j \in J} A_{ij} x_j = b_i, x_J \geq 0.$$

At a solution, we can generate dual variables π on the equality constraints, and determine if there is a dual constraint that is infeasible at π (essentially this is the pricing step of the simplex method). Since the dual constraints are $A^T \pi \le c$, this amounts to finding a $k \notin J$ with $c_k - \pi^T A_{\cdot k} < 0$. If no such k is found, the current solution is optimal, otherwise the additional column k is added to the set J, and further pivoting is done in the simplex method on the restricted master problem. The key point is that instead of looking over all $k \notin J$ to find the new column, in many applications an alternative optimization problem can be solved to *generate* the (best) column. An excellent primer on column generation can be found in [49]. Specific examples of doing this (for multicommodity network flow and cutting stock problems) are detailed in [67, 75]. Dantzig-Wolfe decomposition is a commonly used method for large scale (block structured) linear programs that combines Lagrangian decomposition with column generation. Branch and price [16] extends column generation ideas to the MILP setting.

2.4.4   BENDER'S DECOMPOSITION

Since computational difficulty of optimization problems increases significantly with the number of variables and constraints, solving these smaller problems iteratively can be more efficient than solving a single large problem. Bender's decomposition is a general approach that can be applied to MILP problems of the following form:

$$\min_{x,y} \left(c^T x + d^T y \right) \text{ s.t. } Ax + Hy \ge b, x \in X, y \in Y.$$

Here X represents standard linear constraints on the continuous variables, whereas Y incorporates all the integrality constraints on y as well. Implicitly, for this approach to be successful, it is important that problems with constraint sets X or Y can be solved efficiently, and the constraints $Ax + Hy \ge b$ are complicating restrictions that restrict our ability to optimize easily over x and y. The above problem is equivalent to

$$\min_{y \in Y} d^T y + \left(\min_{x \in X} \{ c^T x \text{ s.t. } Ax \ge b - Hy \} \right). \tag{2.9}$$

For fixed y, the inner minimization is a linear program, and hence by linear programming (weak) duality:

$$\min_{x \in X} \{ c^T x \text{ s.t. } Ax \ge b - Hy \} \ge (b - Hy)^T u$$

for any u that is feasible for the dual problem

$$\max_{u \ge 0} (b - Hy)^T u \text{ s.t. } A^T u \le c. \tag{2.10}$$

Bender's decomposition solves (2.9) by iteratively generating lower bounds (cuts) formed from feasible points of the dual. More precisely, each iteration consists of solving (2.10) for a fixed

$y = \bar{y} \in Y$ or determining an unbounded ray. Then we solve

$$
\begin{aligned}
\min_{z \in \mathbb{R}, y \in Y} \quad & z \\
\text{s.t.} \quad & z \geq d^T y + (b - Hy)^T \bar{u}^k, \quad k = 1, \dots, K; \\
& 0 \geq (b - By)^T \bar{u}^l, \quad l = 1, \dots, L,
\end{aligned}
\tag{2.11}
$$

where k indexes previously found optimal solutions of (2.10) and l indexes previously found unbounded rays of (2.10). The solution $y = \bar{y}$ of this problem is then fed back into (2.10). Note that (2.10) is a linear program, and this method works well when the MILP (2.11) is easier to solve than the original MILP (specifically note that it does not involve the matrix A).

2.4.5 OTHER APPROACHES

There is a recent and growing literature on derivative-free optimization, that is, algorithms for solving (2.4) where f is nonlinear, but for which derivative information is either too expensive to calculate or simply not available. These methods could be used for example to solve the Lagrangian decomposition problem outlined above, or to solve models where $f(x)$ is calculated by a simulation code for given parameters x. The reader is referred to [39, 148] for further information on this topic.

Of course, there are numerous heuristic approaches that can also be used to solve not only MILP problems, but also classes of models mentioned previously. But unlike the methods and codes we outline here, these heuristics produce only feasible solutions for the problem (that is values of x and y that satisfy the constraints of the problem), they do not provide bounds (essentially a certificate) that tell how far the objective function might be from its best possible value.

2.4.6 MODELING LANGUAGES

Optimization technology has improved immensely over the past five decades, and complex implementations of the algorithms are now available. Accessing these solvers, and many of the other algorithms that have been developed over the past three decades, has been made easier by the advent of modeling languages. A modeling language [27, 70] provides a natural, convenient way to represent mathematical programs and provides an interface between a given model and multiple different solvers for its solution. The many advantages of using a modeling language are well known. They typically have efficient automatic procedures to handle vast amounts of data, take advantage of the numerous options for solvers and model types, and can quickly generate a large number of models. For this reason, and the fact that they eliminate many errors that occur without automation, modeling languages are heavily used in practical applications. Although we will use GAMS [33], the system we have used to implement our models, much of what will be said could as well be applied to other algebra-based modeling systems like AIMMS [26], AMPL [71], CVX [82], MOSEL [46], MPL [128], OPL [168], PYOMO [89], and TOMLAB [167].

Each of the different systems has pros and cons, and provides different functionality. For example, AMPL has specialized syntax for piecewise-linear modeling, whereas AIMMS has techniques for GUI development and embedding within other programs. OPL facilitates some of the constraint logic programming extensions outlined above. PYOMO is an open source code, and GAMS provides access to the open source codes contained in the COIN-OR collection [120], it also provides data exchange interfaces to Excel, Matlab, and R, whereas TOMLAB is a dedicated suite of optimization tools within Matlab. The principal MILP solvers (CPLEX, GUROBI, XPRESS) also provide APIs to common programming languages.

In the context of convex programming, the cvx system that is implemented in Matlab follows a different approach. It limits the problem classes to the convex setting, but uses the novel approach of allowing (in fact requiring) the modeler to build up a convex model from given convex functions and providing techniques for composition that preserve convexity. The book [31] provides an overview of the compositional approaches that work in this setting, and there is documentation of the modeling system in [82]. Other modeling constructs allow a modeler to treat uncertainty using stochastic and robust optimization.

2.5 CONCLUSION

Optimization can facilitate prediction, improve operation, and help with strategic behavior and design. Models can be combined; their utility stems from engaging groups in a decision, actually making complex decisions, and operating or controlling a system of interacting parts. Putting together some or all of the constructs that we outlined above requires skill, understanding, and many iterations. Modeling and optimization is best done as part of an iterative process that engages the designer, facilitates the collection of appropriate data, and allows domain-specific design tools to be incorporated into a general (optimization) framework. Determining what is the appropriate model takes some effort: is it linear or nonlinear, deterministic or probabilistic, discrete or continuous, best modeled using smooth or nonsmooth functions, and/or static or dynamic. Modeling systems allow one to move between these formulations quickly, provide some data and model verification, and provide constructs to broaden the modeling classes and tricks that are practically usable.

The following four chapters provide case studies in using mathematical optimization techniques to model the design or operation of various systems. These case studies demonstrate the usefulness of the modeling techniques described in this chapter, the broad range of applicable problems, and practical usefulness of MILP formulations and solvers. The avid reader may wish to skip to the final chapter, conclusions, where we provide insight and experience into deciding which types of problems are suitable for MILP, as well as tuning strategies for improving performance and tips on how to formulate good models.

CHAPTER 3

Case Study: Instruction Set Customization

3.1 INTRODUCTION

We begin our set of case studies with a problem inside the domain of hardware/software partitioning, where a program's execution is split into hardware and software components with the objective of minimizing execution time. Specifically, we explore the task of generating candidate instructions for extending custom processors, and build a model which can be applied in an architecture-generation toolchain. By studying this problem, we seek to elucidate the process and nuances of modeling in Mixed Integer Linear Programming (MILP), and we start by explaining the practice of employing custom instruction sets.

Custom Instruction Sets   In the embedded domain, Application Specific Instruction-set Processors (ASIPs) provide an opportunity for highly efficient execution by tailoring the instruction set to a particular application. Languages like the Language for Instruction Set Architectures (LISA) [93] and Tensilica Instruction Extension (TIE) [79] allow for the manual creation of specific instructions for custom processors. To further automate this process, Tensilica's "Xpres compiler" and Synopsys's "Processor Designer" take as input an application written in C, and produce highly customized processors with application-specific instructions.

These automatic techniques require program analysis and profiling to identify computationally intensive regions of code which can be profitably partitioned into hardware and software components, and be correctly interfaced with the rest of the processor. Typically, this process is broken into two phases: 1) instruction template generation, where candidate instructions are generated based on program properties and estimated usefulness; 2) instruction template selection, where specific instructions are chosen based on the frequency of execution and instruction hardware implementation with associated power, area, and performance trade-offs. In this work, we focus on the template generation portion of the problem.

Instruction Template Generation   Generating custom instruction candidates from an application is the process of analyzing the intermediate representation of the program, and finding subgraphs which can be made into template instructions. Depending on the target system, these candidates must have certain properties in terms of legality (including graph structure or operation types), and should capture a piece of computation in a way that is expected to improve the performance as much as possible with respect to the instruction-by-instruction execution. The problem of instruction template generation naturally resembles Graph Partitioning and Graph Clustering problems, which attempt to divide a graph into subgraphs, where the interface between these subgraphs is minimized. While they could be applicable to some extent, they either lack the ability to optimize for characteristics of the graph which are not part of the graph structure, or cannot provide optimality/legality guarantees on the divisions. For this reason, we explore a mathematical optimization-based-approach. Our approach and treatment of the problem is closely related to that of Atasu [143]. It differs somewhat in the set of constraints used for the model, and our evaluation uses different benchmarks and metrics.

3.2   OVERVIEW

Our high level strategy is to apply integer linear programming to find opportunistic regions of code to accelerate, but target only single basic blocks at a time. This model would be used as part of a design flow, as depicted in Figure 3.1. First, a compiler "frontend" will output the program's basic blocks to the template generator. The template generator uses the MILP model we develop to analyze these basic blocks, and discover templates which are good candidates for instruction extensions. During this phase, a separate graph isomorphism pass eliminates redundant templates. These templates are synthesized into hardware to determine their area and latency. The template selection phase uses this information, and performs an analysis considering the area, coverage, and speedup trade-offs to determine the set of chosen instructions. Finally, the compiler "backend" generates meta-data regarding instruction formats used to re-compile the code for the newly extended instruction set.

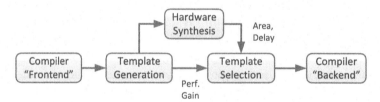

Figure 3.1: Design flow for custom instruction generation and compilation.

System Architecture   The baseline architecture we target, depicted in Figure 3.2, is that of an in-order processor with some number of custom instruction hardware components for performing specialized instructions. The interface to the hardware is the bus which can deliver or receive a certain number of inputs per cycle. As our target does not support memory inputs, we only provide an interface with the register file. It should further be noted that all of the inputs for a custom instruction must arrive before the computation begins, meaning that no intermediate output values may be used in the processor before being returned to the custom instruction hardware.

Example Template   To elucidate the fundamental issues of the problem, we show some example templates in Figure 3.3. Here, circles represent instructions to be computed, and edges repre-

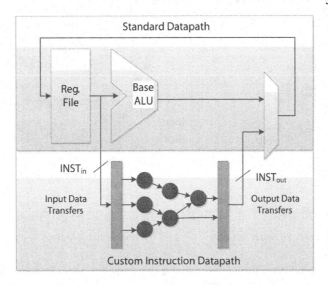

Figure 3.2: System architecture for processor with custom instructions.

sent data dependence. Subfigures (a) and (b) show two different possible instruction templates in shaded colors, each containing different operations and differing numbers of I/Os. These choices affect the potential speedup improvements because certain operations can be accelerated more opportunistically in hardware, and the subgraph chosen will affect the amount of time required to transfer inputs and outputs from the register file. Note that both are "convex[1]" in the sense that no edge leaves and returns to the subgraph. This property is required for correctness, as the generated template must be serializable with the original instruction stream to be usable.

Problem Statement

> **Determine the convex set of operations, forming a custom instruction template, which maximizes the expected latency reduction for the given basic block.**

Chapter Organization   We describe the modeling of the integer linear program by first describing the system abstractly, then writing logical constraints, and where required, linearizing these constraints to match integer linear programming theory. The goal of the next three sections is to describe, from a modeler's perspective, how to formulate the model in terms of the fundamental decision variables, describe the formulation of the constraints by linearizing logical constraints, and finally how to reason about and write the objective. We then describe the limitations of our model, the related work, and an evaluation with a modified compiler and real workloads.

---

[1] Not to be confused with convexity in mathematical optimization theory.

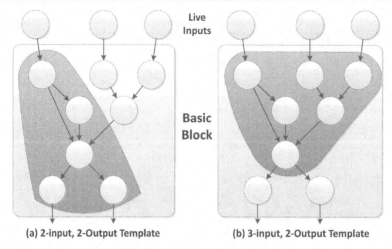

Figure 3.3: Example instruction templates. (a) and (b) show two ways to partition the basic block's instructions, each including different numbers of I/Os.

3.3   FORMULATION: PARAMETERS AND DECISION VARIABLES

The first step of formulating the model is choosing what will be the relevant decision variables, and what are the system parameters needed to capture its solution space. In this section, we also list some of the auxiliary variables that will help to formulate the linear constraints.

Decision Variables   We must first represent the DAG of instructions in our basic block. For this, we define the set V, which represents the constituent instructions, along with instructions outside the basic block which produce values used inside. The connections, or data dependencies, between instructions in V can be described with $\mathcal{A}_{ij}$, the appropriate adjacency matrix. Now the choice of how to represent the subgraph becomes obvious; we associate with each vertex i a variable x_i, which describes if the vertex is contained in the template. This is our decision variable, as it is the output of model we are trying to determine. In fact, all other quantities for this model can be easily computed offline based on these variables.

V Set of instructions in the basic block.

$\mathcal{A}_{ij}$ Node adjacency matrix, which describes connections of node i to node j.

x_i Decision variable indicating if node $i \subset V$ is in the best instruction template.

System Parameters   To model our system, we need to present, as input, the parameters which describe the system, and use these in formulating the model. First, we need to understand the difference between the software execution time of an instruction, and the hardware execution time, so we approximate these with s_i and h_i respectively. In addition to the execution latencies, we need to calculate the data transfer time for live inputs and outputs. This is based on the bandwidth and latency to the register file.

These are used to calculate the number of data transfer cycles C, which should be minimized if possible. The induced latency is parametrized by the variables $PORT_{in}$ and $PORT_{out}$, which describe the number of register file read/write ports, and RFC_{in} and RFC_{out}, the number of cycles to access the register file. This is captured by $PORT_{in}$ and $PORT_{out}$ for bandwidth, and RFC_{in} and RFC_{out} for latency.

s_i, h_i Estimated software, hardware cost of node i.

$PORT_{in}$, $PORT_{out}$ Number of input, output register file ports.

RFC_{in}, RFC_{out} Number of cycles to read, write to the register file.

Auxiliary Variables   The overall expected latency reduction, termed Z, is the combination of three terms: the hardware latency for the template H, the software latency of the template S, and the data transfer time C. We also need variables to enforce the legality of the template.

The hardware time H is approximated as the critical path through the template as if it was implemented in hardware. We need a new variable here to represent the execution time of each instruction in the template, t_i. The software latency, because we have a simple in-order machine, is approximated as the sum of the latencies of the instructions, and we need no further variables to describe this.

To compute the data transfer time for the chosen template, we'll need to use the graph structure to determine the number of inputs and outputs, $INST_{in}$ and $INST_{out}$ respectively. We use these values to compute the number of data transfers DT_{in} and DT_{out}.

Finally, we must model the restriction that the template is one continuous region of the instruction graph, a property which we defined earlier as "convexity". We can enforce this by noticing that no non-template member should have an ancestor and descendant inside the template. We will calculate these relationships for every node using a_i and d_i, for determining if an ancestor and descendant is inside the template respectively.

Z Estimated Latency Reduction for Template.

H Estimated Hardware Cycles to execute template.

S Estimated Software Cycles to execute template.

$INST_{in}$, $INST_{out}$ Number of template inputs, outputs.

DT_{in}, DT_{out} Number of register file data transfers in, out.

t_i Cycle time for Node i in template.

a_i, d_i Ancestor or Descendant is in the template.

Input Parameters: Software Description	
V	Set of computation vertices.
s_i, h_i	Estimated software, hardware cost of node i.
$\mathcal{A}_{ij} \subset V \times V$	Arcs Between Vertices.
Input Parameters: Architecture Description	
$PORT_{in}, PORT_{out}$	Number of register file read/write ports.
RFC_{in}, RFC_{out}	Cycles to access register file.
Variables: Final Outputs	
x_i	Node i is in the template.
Z	Estimated Latency Reduction for Template
Variables: Intermediates	
$input_i, output_i$	Node i is in the input or output interface.
$INST_{in}, INST_{out}$	Number of inputs & outputs.
DT_{in}, DT_{out}	Data transfers in & out.
C	Total cycles to transfer inputs/outputs.
t_i	Cycle time for Node i in template.
H, S	Estimated Hardware & Software Cycles to execute template.
a_i, d_i	Ancestor or Descendant is in the template.

Table 3.1: Summary of formal notation used

Table 4.1 summarizes the parameters and variables of the model, in terms of input parameters of the architecture and input basic block graph, output decision variables, and intermediate variables that describe specific aspects of the problem. Any variables introduced for the linearizations of constraints are also included in this table.

3.4   FORMULATION: CONSTRAINTS

For simplicity of explanation, we describe the model in two parts. First, we show how to model determining the number of inputs and outputs, in order to account for the transfer time between the register file and the custom hardware. Then, we show how to model the convexity requirements, which enforces that no value can leave and return to the main hardware.

Inputs and Outputs   We can determine the number of inputs and outputs using the following observation. An input to the graph is one such that, for a given edge $(i, j) \in \mathcal{A}_{ij}$, i is not inside the template, but j is. The reverse case is true for outputs. The logical formulation of the number of inputs and outputs is therefore as follows.

$$INST_{in} = \sum_{i \in V} \neg x_i \wedge \left(\bigvee_{j:(i,j) \in \mathcal{A}} x_j \right)$$

$$INST_{out} = \sum_{i \in V} x_i \wedge \left(\bigvee_{j:(i,j) \in \mathcal{A}} \neg x_j \right)$$

Since MILP does not operate on logical predicates directly, we must linearize the equations using the techniques described in Chapter 2, Section 2.3.1. By distributing the logical and over the conjunction in the equations above, and by introducing an auxiliary variable for each equation $input_i$ and $output_i$, we can now enforce the implications $\neg x_i \wedge x_j \implies input_i$ and $x_i \wedge \neg x_j \implies output_i$ with the following linear equations:

$$input_i \geq (1 - x_i) + x_j - 1 \quad \text{for all } (i, j) \in \mathcal{A} \tag{3.1}$$
$$output_i \geq x_i + (1 - x_j) - 1 \quad \text{for all } (i, j) \in \mathcal{A} \tag{3.2}$$

Now, we can simply use the following two equations to add up the total number of inputs and outputs.

$$INST_{in} = \sum_{i \in V} input_i \tag{3.3}$$

$$INST_{out} = \sum_{i \in V} output_i \tag{3.4}$$

If required, at this point we could enforce the number of inputs or outputs to be at certain values by introducing appropriate equations. For example, we may only want to find 2-input, 1-output instructions, if this is what is encodable in the instruction. However, for the rest of our formulation and analysis, we will assume the hardware can transfer additional values from the register file.

Convexity   As mentioned earlier, we enforce the property of convexity as a subgraph where no inputs are dependent on outputs, essentially creating a single continuous graph (although independent subgraphs can exist inside the template). We can enforce this property by first calculating template ancestor and descendant relationships as follows. Notice that any node which has a parent which has an ancestor in the template also has an ancestor in the template, and likewise for descendant and child relationships. This leads to a logical formulation for ancestor and descendant of each node like the following.

$$a_j = \bigvee_{j:(i,j)\in\mathcal{A}} (x_j \vee a_i) \quad \text{for all } j \in V$$

$$d_i = \bigvee_{i:(i,j)\in\mathcal{A}} (x_i \vee d_j) \quad \text{for all } i \in V$$

These equations can be linearized by simply using the identity for $x_i \vee a_i \implies a_i$ discussed in the previous chapter. The result is the following two equations.

$$a_j \geq x_j + a_i \quad \text{for all } (i, j) \in \mathcal{A} \tag{3.5}$$

$$d_i \geq x_i + d_j \quad \text{for all } (i, j) \in \mathcal{A} \tag{3.6}$$

Finally, to enforce the convexity constraint, we just need to assert that no non-template node should have an ancestor and descendant in the template. We can do this by taking the logical conjunction of the three associated variables.

$$\neg(a_i \wedge d_i \wedge \neg x_i) \text{ for all } (i) \in V$$

This can be linearized by combining identities from Table 2.3 on page 35 in the previous chapter.

$$d_i + a_i + (1 - x_i) \leq 2 \quad \text{for all } (i) \in V \tag{3.7}$$

3.5   FORMULATION: OBJECTIVE

Now that we have enforced the appropriate graph properties, we need to determine the efficacy of the template in terms of reduced latency. In other words, we must form our optimization objective.

The first step is to compute the total expected time for the template to be executed in software, which, for an in-order processor, can be computed as the following.

$$S = \sum_{i \in V} x_i * s_i \tag{3.8}$$

Next, we must consider the custom instruction latency if it were actualized in hardware. This is estimated by calculating the critical path through the hardware nodes. For each edge in the template, we simply add the source node's expected time, and the estimated hardware latency of the current operation. The result is accumulated in the auxiliary continuous variable t_j. This naturally yields a linear equation. The subsequent equation simply takes the maximum time as the hardware latency.

$$t_j \geq t_i + h_i * x_i \quad \text{for all } (i, j) \in \mathcal{A} \tag{3.9}$$

$$H \geq t_i \quad \text{for all } i \in V \tag{3.10}$$

The remaining latency to be calculated is in the transferring of data from the standard datapath to the custom instruction. If the number of operands is greater than the number of input or output ports of the register file, we add additional data transfers DT_{in} and DT_{out}. These can be calculated simply as the number of instructions divided by the number of ports, subtracting one for the first set of I/Os which are pipelined. Though this is a continuous number, the DT variables are integer, so they are appropriately rounded.

$$DT_{in} \geq INST_{in}/PORT_{in} - 1 \tag{3.11}$$

$$DT_{out} \geq INST_{out}/PORT_{out} - 1 \tag{3.12}$$

The total transfer time C takes into account the number of data transfers, as well as the cost of each transfer.

$$C = DT_{in} * RFC_{in} + DT_{out} * RFC_{out} \tag{3.13}$$

Finally, the total reduction in cycles can be written simply as the difference between the total software latency, and the sum of the hardware and data transfer latencies.

$$Z = S - (H + C) \tag{3.14}$$

3.6   MODELING LIMITATIONS

To formulate the problem as an efficient MILP, some trade-offs were made in the modeling. We describe some of the limitations below.

Hardware Latency   Here we considered the hardware latency to be the critical path through estimates of each hardware unit in the template. This is not necessarily accurate, because the actual synthesis will have a different latency. Synthesizing every combination in the optimizer is impractical, and since the equations involved are non-linear, would not fit into the linear programming paradigm.

Software Latency   The software latency was calculated as the sum of each of the software latencies of individual instructions. This ignores pipeline effects like data dependence hazards or dynamic resource conflicts. However, to the first order, it should be accurate enough for our baseline processor to be useful.

Basic Block Scope   We examined only non-control flow instructions for inclusion in custom instructions, and only examined one basic block at a time. This made the scope of the problem more tractable. That said, others have still managed to use MILP for performing custom instruction generation across larger program regions [73].

Iterative Solutions   Each time we find a solution, we fix the nodes in the template to not be in the template, to find additional potential subgraphs inside the basic block. This way of iterating over the space means that we will not find every possible template, because there could have been useful overlapping templates that our strategy does not consider.

3.7 EVALUATION

In this section, we answer two important questions:

1. Can this model be applied in the context of a real system, and be implementable?
2. Is the problem formulated well enough to be solved in a reasonable amount of time?

3.7.1 METHODOLOGY

We wrote the model described in this chapter in GAMS, using roughly fifty lines of code. In order to evaluate the model on real workloads, we integrated the GAMS model into an LLVM module which analyzes data dependence inside basic blocks. The two components communicate through graphs written in GAMS syntax, which include all of the input parameters from Table 3.1.

For this particular problem, we consider arithmetic and logical operations as candidates for specialization, and all other instructions are prevented from inclusion in the templates by fixing their template decision variables to zero. For this problem, we've set the register file access cost (RFC_{in} and RFC_{out}) to a single cycle, and give two register read ports ($PORT_{in}$) and one write port ($PORT_{out}$).

3.7.2 RESULTS

Implementability   We apply our scheme to the EEBMC benchmark suite, and consider basic blocks which have more than one computation instruction which would be feasible to include in a custom instruction template. A summary of our results appears in Table 3.2. We characterize the workloads with the number of basic blocks and average basic block size, appearing in the second and third columns of Table 3.2. Overall, we see a significant number of basic blocks for potential specialization.

Result-1: Our model for template generation is practical and implementable.

Solution Time   The fourth and fifth columns of Table 3.2 show the average number of equations and variables for each problem instance, which are very small due to small average basic block size. This leads to very fast solution times, shown in the final column. We show the expected latency

Benchmark	Num BBs Considered	Average BB Size	Model Eqs	Model Vars	Per Temp. Lat. Reduct.	Per Temp. Solve Time(ms)
a2time01	272	9.98	248	126	3.94	89
aifftr01	232	11.54	338	167	5.33	95
aifirf01	238	11.68	321	162	4.85	113
aiifft01	226	11.50	338	167	5.33	215
autcor00	200	10.99	307	153	5.02	204
basefp01	208	11.68	319	158	5.16	234
bezier01	198	11.20	326	161	5.35	149
bitmnp01	349	8.77	304	152	4.80	79
cacheb01	208	11.56	338	168	5.42	75
canrdr01	234	10.88	305	153	4.85	90
cjpeg	1080	10.62	186	104	2.36	77
conven00	209	10.99	315	156	5.14	90
dither01	200	11.19	317	159	5.12	104
djpeg	857	11.67	233	129	2.47	103
fbital00	204	10.82	295	148	4.85	80
fft00	215	11.26	306	153	4.90	70
idctrn01	227	15.97	468	230	10.44	73
iirflt01	251	11.89	351	176	5.72	92
matrix01	302	10.53	311	156	5.05	95
ospf	208	11.07	325	161	5.28	76
pktflow	238	10.76	312	157	4.84	105
pntrch01	221	10.90	320	160	5.16	96
puwmod01	236	10.73	282	143	4.40	194
rotate01	295	9.59	263	135	3.74	96
routelookup	219	10.99	318	158	5.11	76
rspeed01	212	11.16	324	162	5.08	93
tblook01	224	11.60	367	181	5.35	77
text01	215	10.75	323	160	5.28	99
ttsprk01	303	11.80	344	174	4.41	105
viterb00	217	11.65	348	175	5.03	112

Table 3.2: Average number of equations and solve times

reduction (in cycles) of each template in the sixth column, as an average over the entire benchmark. These tend to be in the range of 5-10 cycles on average, but can be as large as 50 or more cycles for certain templates. It is more likely that the large latency reduction templates would be picked up by the instruction selection phase, as they show more benefits. Overall, we see that the model can both be solved in a reasonable amount of time, showing its feasibility, and can find templates with large expected latency reduction, showing its effectiveness.

Result-2: The formulation is solvable on a reasonable time scale.

3.8   RELATED WORK

For a comprehensive survey of techniques, practices, and research in the field of automatic instruction-set extensions, Galuzzi et al. provide a comprehensive survey [72]. We highlight the differences between the MILP approach explored in this chapter and a few selected approaches below.

Many practical techniques have been explored to solve the problem of automatic instruction set extension. These algorithms come in a variety of flavors, and must trade-off the execution time of the algorithm, the scope of search inside the program structure, and the type of graphs searched for. Hence, these solutions sacrifice exploring some aspect of the design space. Some solutions only look for connected subgraphs, which ignore potential increased parallelism [15, 183]. Other techniques only find instruction templates with certain simplified graph structure, like those that only find single-output subgraphs [8].

The MILP approach we explore here attempts to generalize the type of graphs we look for, including those with multiple outputs, useful for computationally intense code, and even those which have disjoint subgraphs (yet still convex), as they provide additional parallelism. Our approach guarantees generality in subgraph form, and also in optimality, given that the latency model can be configured to match the hardware.

3.9   CONCLUSIONS

In this chapter, we identified a problem suitable for integer linear programming, custom instruction set generation, and described how to create the formal model. We described the modeling process, from high level description, to integer linear constraints. Integrating our solution into a compiler, we were able to generate instruction templates with optimal projected performance improvement. Our implementation proved to be efficient, taking on the order of 100ms per solution.

CHAPTER 4

Case Study: Data Center Resource Management

4.1 INTRODUCTION

In this chapter we describe how to apply MILP modeling in an entirely different domain in computer architecture, specifically in large scale computing. Today's data centers house vast numbers of machines and must support an equally large quantity of services. By "service", we mean an instance of a particular workload for one machine. However, these services do not typically fully utilize the underlying machine at all times, and therefore uninformed service allocation can lead to poor utilization. A typical strategy for dealing with poor utilization is to co-locate instances of multiple services, either in the form of short-term jobs or services hosted on virtual machines, onto the same physical machines, while retaining the equivalent quality of service (QoS).

Data-center resource managers are centralized software components which manage the execution of a large number of services. Depending on the role of the resource manager, the granularity of allocation can be performed at different levels, including globally and statically for all services, periodically for a certain time window, or incrementally as each job or set of jobs is requested. The longer the time window we choose to allocate for at once, the better a solution we can and should seek to attain.

Since co-locating multiple services on the machine could degrade performance, it is critical that the resource manager effectively allocate machine resources. To perform this task adequately, the resource manager needs some information about services and machines which can facilitate the mapping. For example, if the resource manager has information about a service's expected CPU and memory and network bandwidth requirements, and if it knew the respective capacities of a machine, it could allocate services to machines in such a way that they would likely not interfere, at least in terms of the modeled resources. Figure 4.1 shows an example of this problem.

Not only could the inputs vary, but the ultimate goal of the resource manager could be different depending on the setting. For instance, one goal may be to use the least number of machines, so that other machines could be powered off. A more nuanced goal minimizes the overall power consumption. Further, if the jobs are commoditized, the overall goal could be to maximize the expected revenue, where quality of service could be degraded at the cost of reduced price per job.

This chapter will explore a number of related problems, applying MILP to the problem of data center resource management. We create one basic model, then build on top of this to

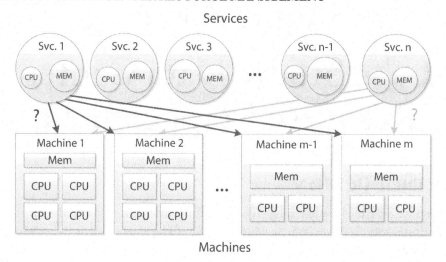

Figure 4.1: Example resource management system with heterogeneous hardware and software.

capture the constraints or requirements of different situations. Our treatment of this problem is most closely related to that of Speitkamp and Bichler, who formulate the static problem in MILP, and evaluate their model on real workload data [114]. By implementing our models and measuring the run-times and optimization bounds, we will explore how they can be applied in real world scenarios, and how simple models can be extended to model complex requirements and phenomena.

4.2 OVERVIEW

Static Server Allocation (SSAP)   Essentially, the problem is to allocate services with fixed resource requirements onto machines with fixed resource limitations. The problem is static in nature, in that we don't take advantage of the ability to migrate jobs across machines, and assume that allocations are final. For this problem, we also assume the most general case, where machines are allowed to provide heterogeneous resource requirements, and all instances of services have been profiled individually. Computationally, this problem is equivalent to the multidimensional vector bin-packing algorithm, where each tuple of resources corresponds to a vector.

Warehouse Server Allocation (WSAP)   The underlying assumption of the previous problem, SSAP, is that each workload is unique. This is often not the case in warehouse settings, where there will likely exist many workloads of a particular type. Classifying workloads can have many benefits, the first being that we can better characterize them through aggregated profiling, etc. Indeed, Vasić et al. give a methodology for classifying virtual instances into a small number of workload classes [170] We can also take advantage of this information when performing the allocation, to improve the formulation and reduce the computational complexity. We call this the Warehouse Server Allocation Problem (WSAP).

Time-Varying Server Allocation (TSAP)   In certain situations, we may have *a priori* knowledge of the resource usage pattern of a workload in the course of a day. For instance, certain services may experience more load during the nighttime, or daytime. Therefore, since we are still considering static allocation, we may want to co-locate machines which have complementary resource patterns, ensuring that we do not exceed the machine's resource limitations. Figure 4.2 shows two example services with an opportunistically co-locatable CPU resource utilization pattern. This is termed the Time-varying Server Allocation Problem (TSAP).

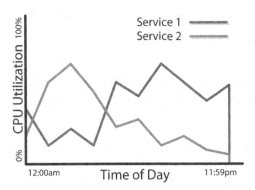

Figure 4.2: Two services with complimentary resource patterns.

Interference-Sensitive Server Allocation (ISAP)   Although we have modeled the machine resource usage, it is true that the performance of co-located jobs can be affected by interference in the memory system. This is especially true for latency sensitive tasks, where cache interference and contention for memory bandwidth can have a profound effect on the overall Quality of Service (QoS). Moreover, these types of resource constraints do not easily lend themselves to the "bin-packing" approach we have thus far proposed to manage system resources, as there is a complex relationship between memory system contention and performance degradation, and this relationship is specific to the applications involved.

To overcome these problems, Mars et al. show how we can capture this relationship with the concepts of memory pressure and memory sensitivity [126]. Here, our abstraction for *memory pressure* is the application's lowest level working set size which determines the amount of on-chip cache used for storing the data for the given application. The *memory sensitivity* of an application describes the amount of on-chip cache required by the application to achieve a given quality of service. Figure 4.3 gives a concrete example of the relationship between memory pressure (from all other applications on the machine) and the projected application QoS. In the given example, to maintain a quality of service higher than 90%, service 1 and service 2 require a system memory pressure of less than 10MB and 25MB respectively. Because they each exert less than the other's

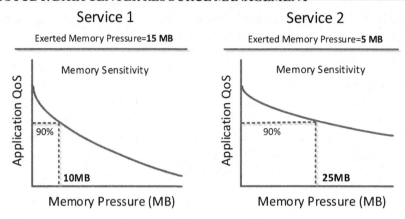

Figure 4.3: Memory pressure and sensitivity for two complimentary services.

memory sensitivity limit (15MB≤25MB and 5MB≤10MB), they can be co-located effectively. The problem which incorporates the constraint that QoS should not fall below a certain threshold is called the Interference-sensitive Server Allocation Problem (ISAP).

Problem Statement   We have described several problem classes above for data center resource management. Though the requirements and objectives for each model described above is slightly different, the overall nature is the same. We can state the overall problem as:

> **Statically determine the best co-locations of services on servers such that the resource requirements and service level agreements can be satisfied.**

Chapter Organization   Similar to the previous chapter, the next three sections describe the modeling of the integer linear program by first describing the decision variables and input parameters, then formulating the mixed integer linear program through linearizations of logical constraints and the objective function. We then discuss modeling limitations and related work, and conclude with an evaluation.

4.3   FORMULATION: PARAMETERS AND DECISION VARIABLES

We will describe the models for data-center resource management following the structure of the previous chapter. In this section, we will decide the relevant decision variables, and determine what inputs are needed to solve the problem. The following sections describe the MILP model formulation.

Deciding Decision Variables   The essence of our problems is to find the allocation, or "mapping" between a set of services S and a number of machines, or compute resources C. This mapping we call $M_{SC} \subset S \times C$. Depending on the problem, the mapping can be a binary variable representing a particular mapping of a service onto a machine, or it can be an integer variable, representing the number of services of a particular type mapped to a machine.

S Set of Services

C Set of Compute Resources/Machines

M_{SC} Mapping from Services to Machines

System Parameters   Both services and machines require a notion of resources, K, which is a set that could include CPU utilization, memory utilization, network bandwidth utilization, or any other shared resource. The required resources for the services is defined by $R_{SK} \in \mathbb{R}^{|S \times K|}$, and the resource limitations for the machines is $L_{CK} \in \mathbb{R}^{|C \times K|}$ Since we also model time-varying resources, we add a new set T for the time intervals, and add a time dependent resource requirement variable $R_{SKT} \in \mathbb{R}^{|S \times K \times T|}$

We also need parameters for memory pressure P_S and memory sensitivity X_S. Memory pressure is really a measure of workload size, and memory sensitivity is the memory pressure which can be tolerated for some quality of service.

K Set of resources, like CPU and Memory

R_{SK} Resource requirements of service S of type K

L_{CK} Resource limitations of machine C of type K

T Set of time intervals

R_{SKT} Resource requirements of service S of type K, at time T

P_S Memory pressure for service S

X_S Memory sensitivity for service S (pressure of system tolerated for QoS)

Auxiliary Variables   We'll need a couple of auxiliary variables to optimize the mapping. First, we need to know which machines themselves are on, $O(C)$. We also define our objective variable, TOT_{on}, the total number of machines which are required to satisfy all the services' requirements.

$O(C)$ Binary variable indicates whether machine C is on.

TOT_{on} Total number of machines which are on.

Table 3.1 summarizes the parameters and variables of the model in terms of parameters related to the services and machines, output decision variables, and intermediate variables.

4.4   FORMULATION: CONSTRAINTS

We now describe the formulation of the four problems we model in this chapter. We begin with the simplest problem, and build off of this problem by adding and modifying constraints.

	Input Parameters: Services and Requirements
S	Set of Services.
R_{SK}	Resource requirements of service S of type K.
T	Set of time intervals.
R_{SKT}	Resource requirements of service S of type K, at time T.
P_S	Memory pressure for service S.
X_S	Memory sensitivity for service S (pressure of system tolerated for QoS).
	Input Parameters: Machine Resources
C	Set of Compute Resources/Machines.
L_{CK}	Resource limitations of machine C of type K.
	Variables: Final Outputs
M_{SC}	Mapping from Services to Machines.
TOT_{on}	Total number of machines which are on.
	Variables: Intermediates
$O(C)$	Binary variable indicates whether machine C is on.

Table 4.1: Summary of formal notation used

Static Server Allocation (SSAP)   The most basic constraint that we need to enforce is for there to exist a mapping from services to machines, such that each service is mapped to some machine. We enforce this simply with the following equation, which forces only one M_{sc} to be valid for any c.

$$\sum_{c \in C} M_{sc} = 1 \quad \text{for all } s \in S \tag{4.1}$$

Recall that for SSAP, each server and machine is unique, and that we need to enforce the resource limitations, and we need to determine which servers are "on". The first constraint below, a linear constraint, enforces proper utilization. For each resource and machine, we add up the contribution of resource usage from each mapped job, and limit that against the max for that machine. The second equation below, a logical constraint, equates the fact that no service is mapped with the fact that the machine is off.

$$\sum_{s \in S} (M_{sc} * R_{sk}) \leq L_{ck} \quad \text{for all } c \in C, k \in K$$
$$\bigwedge_{s \in S} (M_{sc} = 0) = \neg O_c \quad \text{for all } c \in C$$

We can actually linearize the above two constraints as the following single constraint, by simply multiplying the right hand side of the first constraint above by O_c. The reason this works is that if any M_{sc} is on, then O_c must be on to enforce the limit. Since we will later see that our objective is to minimize the sum of O_c, we don't need to enforce the implication the other way.

$$\sum_{s \in S} (M_{sc} * R_{sk}) \leq O_c * L_{ck} \quad \text{for all } c \in C, k \in K \tag{4.2}$$

Warehouse Server Allocation (WSAP)   The warehouse server allocation problem is similar, except that we take advantage of the limited number of workload types. The set S now represents these types, and M_{sc} now represents the *number* of services of s type mapped to machine c. We add one component to the model, num, which indicates how many of a certain type we have. The formulation must be modified so that we can allow multiple types of a certain service to be mapped to the same machine. Constraint 4.3 below accomplishes this, and we borrow the same Equation 4.2 for this formulation.

$$\sum_{c \in C} M_{sc} = num_s \quad \text{for all } s \in S \tag{4.3}$$

For WSAP, we will also assume that we have homogeneous hardware, because this is usually the case (at least to some extent) in the warehouse computing domain. This opens an opportunity for reducing the search space of the problem. Since each machine is the same, it is wasteful to explore allocating services on all available combinations. For instance, if we have 100 machines total, and the optimal solution uses 50 machines, then there are $\binom{100}{50}$ optimal solutions. We can help reduce the symmetry of the problem by considering the early nodes first. We accomplish this by specifying that each machine can only be on if the previous machine is on, which we refer to as canonical order.

$$O_{c1} \geq O_{c2} \quad \text{for all } c_1 \in C, c_2 \in C \text{ s.t. } c_2 \text{ follows } c_1 \tag{4.4}$$

Note that the above principle can be applied simply to limited heterogeneity with the same benefits.

Time-Varying Server Allocation (TSAP)   To take into account time-varying resource requirements, as defined by the parameter R_{skt}, we can make a simple change to the utilization equation. Here, we just enforce the resource limitation for every time period.

$$\sum_{s \in S} (M_{sc} * R_{skt}) \leq O_c * L_{ck} \quad \text{for all } c \in C, k \in K, t \in T \tag{4.5}$$

Here we can again borrow constraints 4.3 and 4.4.

Interference-Sensitive Server Allocation (ISAP) To model memory interference-sensitive workloads, we leverage the memory sensitivity X_S, and memory pressure P_S. Logically, what we need to say is that if a particular kind of service is mapped, then the memory pressure of all other co-located services must not exceed the memory sensitivity for this service. First we need a new variable $Y(S, C)$, to indicate that a type of service is mapped. Then the constraints become the following.

$$M_{sc} >= 1 \implies Y_{sc} \qquad \text{for all } s \in S, c \in C$$
$$Y_{s_1c} \implies \sum_{s_2 \in S} M_{s_2c} * P_{s_2} - P_{s_1} \le X_{S_1} \quad \text{for all } s_1 \in S, c \in C$$

We can linearize these using theorem 2.1 from Chapter 2.

$$M_{sc} \le bigM_1 * Y_{sc} \qquad \text{for all } s \in S, c \in C \tag{4.6}$$
$$\sum_{s_2 \in S} \left(M_{s_2c} * P_{s_2c} \right) - P_{s_1} - bigM_2 * (1 - Y_{s_1c}) \le X_{S_1} \quad \text{for all } s_1 \in S, c \in C \tag{4.7}$$

Though the above equations are correct, they contain arbitrarily large constants which can reduce the effectiveness of the model, and increase the solution time. We should instead limit bigM's to the smallest legal value. The smallest legal value for $bigM_1$ is the most services S that can fit on the machine C. We can compute this simply offline. The smallest legal value for $bigM_2$ is the most memory pressure that a machine could ever take. Here, we use the biggest memory sensitivity given in the problem (X_S).

Finally, we can again borrow constraints 4.3 and 4.4 to complete the TSAP model.

4.5   FORMULATION: OBJECTIVE

The objective for all of the above problems is the same, minimize the number of required machines. We calculate the objective as:

$$TOT_{on} = \sum_{c \in C} O_c \tag{4.8}$$

4.6   MODELING LIMITATIONS

The formulations explored in this chapter have only scratched the surface of the types of features, requirements, or objectives that could be useful for the data center resource manager. Thus, there are many formulations which could yet be explored. We outline a few categories of these below.

Scheduled Migrations   It may be possible, in certain settings, to migrate services across servers as a way of responding to known changes in demand. The cost of migration could be an input to the model, and we could schedule migrations in advance to adapt. This would be a straightforward extension to the model.

Dynamic Resource Requirements   If service resource requirements dynamically change rapidly, it may be difficult to predict the requirements very far into the future, making the SSAP an unattractive formulation. We could, instead, follow the work of [30], which repeatedly makes decisions at intervals to determine the best allocation given the current allocation and current expected demand. Our results show that this would be possible, as scheduling a single time period can be done in a short enough amount of time.

Communication Modeling   In our model, we have not considered, to any rigorous degree, the communication requirements of the various services, nor do we model the underlying network bandwidth capabilities or latencies between servers. For certain types of services, where communication is the bottleneck, this model would not be appropriate. Though we do not explore communication modeling in the data center setting, the next chapter describes an assignment problem formulation which does communication scheduling, but in a very different domain.

4.7   EVALUATION

In this section, we evaluate the four different models using synthetic workload inputs. The key questions are:

1. Can these models be implemented in GAMS and be solved to some degree of optimality?

2. How well do these models scale, and to what problem domains could they be applied?

3. How important is the formulation of the model?

4.7.1   METHODOLOGY

For the model inputs, we must have knowledge of the services in question, including resource usage over time, memory pressure, and memory sensitivity. Since we are not trying to validate our models against real data-center workloads and hardware, or prove that they can achieve a better degree of co-location over another algorithm, we simply use synthetic inputs. This is reasonable, because our goal is to show what is possible using MILP. Though it is true that the exact resource distribution will affect the solving time of the MILP, we posit that these are second order effects, and that synthetic inputs are sufficient given that they are reasonable.

We generate synthetic inputs as follows. We use two resources for set K, modeling CPU and memory resources. For the set of machines, we normalize the hardware capabilities to unit dimensions, and when modeling heterogeneous hardware, add in a uniform distribution from 0 to 1 units. For services, we use a uniform distribution from 0 to 1, and use 8 time periods when modeling time varying resource requirements. Memory pressure and sensitivity are also given uniform distributions, but sensitivity is given a multiplicative factor (average 3×), so that co-location is possible.

For each problem size, we report the average of five runs with different random seeds. Also, for all experiments, we stop the solution procedure when the incumbent solution is within 10%

optimality, or in other words, when the number of additional machines used is less than 10% of the projected optimal value. Note that this doesn't necessarily mean that we haven't attained the optimal value, just that we haven't proven the incumbent solution is optimal.

For comparison, we implement a very simple first-fit algorithm which attempts to place services onto machines iteratively, in much the same way that a dynamic service allocator would function.

4.7.2   RESULTS

Implementability   We begin the results section by giving a summary of the results for all four models with 50 machines and 50 services to be scheduled, as shown in Table 4.2. The first two columns show the number of equations and variables for the problem. SSAP requires many more variables because in this problem we are mapping each service individually instead of mapping "kinds" of services as in the other problems. TSAP and ISAP require more equations, as we are additionally enforcing resource restrictions for time periods and memory sensitivity respectively. The next two columns show the Heuristic and MILP solution optimality (the percentage difference between the given solution and the lower bound which the solver reports). Here, the MILP solution considerably outperforms the Heuristic, especially for the SSAP problem, where the number of choices are high. The final column of Table 4.2 shows the average solve time for each benchmark, which are all reasonably fast.

Result–1: Our formulation is practical and implementable.

Problem	Avg. # Eqs	Avg. # Vars	Heuristic Optimality	MILP Optimality	Avg. Solve (sec)
SSAP	281	2550	61.9%	4.6%	2.41
WSAP	155	300	15.5%	4.8%	0.09
TSAP	855	300	16.3%	6.0%	0.61
ISAP	655	550	14.0%	3.4%	0.23

Table 4.2: Average number of equations and solve times, for 50 services and 50 machines

Scalability   We explore the scalability of the approach by changing the number of machines and services in our problem. Figure 4.4 shows the scalability of all four problems. As expected, the problems scale to a similar degree, where SSAP takes significantly longer than the three others, as it models each individual service. In fact, for SSAP problems larger than 1000 services/machines, GAMS always takes longer than the given timeout of 20 min to find the initial value to the solution. This is because the problem is simply too large. Figure 4.5 shows the benefits we get with the MILP approach for different problem sizes. The MILP solutions are generally about 50% better than the first fit for the SSAP problem, while the others are around 10% better performing, independent of the problem size itself.

Result-2: Our model for data-center allocation scales up to thousands of machines and services.

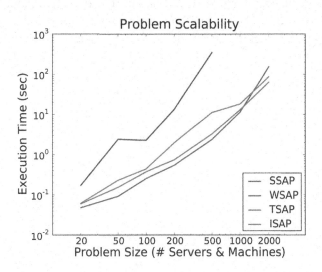

Figure 4.4: Scalability of the four problems.

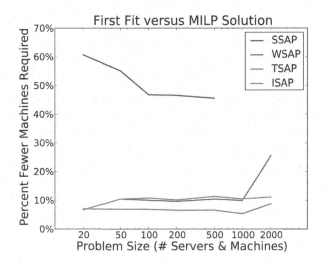

Figure 4.5: Comparison to First-Fit Heuristic Algorithm.

Effects of Model Formulation We now examine the benefits of two optimizations explained in the formulation. The first is regarding symmetry as it relates to which machines are on (O_c) in the WSAP problem. We introduced constraint 4.4 so that we would only consider turning machines on in canonical order, reducing the total number of equivalent solutions by a combinatorial factor. Figure 4.6 shows the benefits of including this constraint by comparing solution times, where "WSAP (SYMM)" is the WSAP problem without the symmetry breaking constraint. Solution times are generally worse, and much more "erratic" as certain instances force the solver to explore many similar solutions.

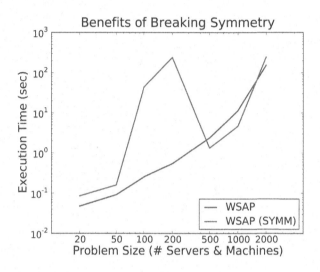

Figure 4.6: Effect of breaking symmetry in the WSAP model.

Another common problem with modeling logical operations in MILP is in choosing the value of so-called "bigM" variables. Those that are too large can create models with very loose linear relaxations, greatly increasing solve time. In Figure 4.7, we show the benefit of using appropriate bigM values by comparing the optimized ISAP with the non-optimized "ISAP-BIGM." This non-optimized formulation uses a nominal value of 1000 for both bigM variables. We see that, with poor bigM values, the problem becomes much more difficult to solve, somewhat increasing the solve times for smaller input sizes.

Result-3: The choice of constraints and parameters in the formulation can significantly affect the solve time.

4.8 RELATED WORK

Research into datacenter resource management spans a vast array of approaches. Interesting examples include using fuzzy logic modeling and inference [181], machine learning techniques [24], and using analytical modeling frameworks with combinatorial optimization [137]. Compared to these types of approaches, the MILP approach may take longer, but can put bounds on the degree

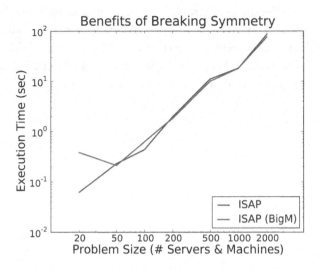

Figure 4.7: Effect of improving bigM in the ISAP model.

of optimality obtained. The value of optimality over shorter runtime depends on the particular resource allocation problem.

Even inside the domain of using Integer Linear Programming for resource allocation, there is still a wide variety of relevant research. As mentioned before, we draw heavily upon the work of Speitkamp and Bichler for their modeling of the SSAP and TSAP problems [158]. Their work also provides extensive studies of real workload data, where we are simply using synthetic workloads. Bose and Sundarrajan use a similar formulation to solve the same problem, but use more nuanced constraints for enforcing service level agreements [30]. Berral et al. and Lubin et al. create models which focus on minimizing energy consumption, which leads to somewhat different constraints [23, 121]. An interesting application of Lubin's MILP allocation model is in the work by Guerva et al. [84]. Here, they use a market-based mechanism to resolve application demands for heterogeneous resources, and the periodically run MILP model clears the market, maximizing the overall welfare of the system. Lastly, Zhu et al. use MILP for making allocation decisions in the data center, but make extensions for modeling the network organization and bandwidth constraints [185].

4.9   CONCLUSIONS

In this chapter, we described how to formulate an important set of related data center problems, specifically the consolidation and allocation of services onto servers, while preserving service level agreements when colocating. We've shown two important phenomena in this chapter: 1. Related problems can be modeled easily through simple extensions to a generic MILP model. Only a few equations were required to model the problems of handling workload and machine characteristics found in warehouse computing (WSAP), time-varying service resource requirements (TSAP), and managing memory interference between services (ISAP). 2. Large MILP problems can be modeled, but sufficient abstractions and efficient formulations must be applied. Here, we showed how tight bigM bounds and reducing symmetry were extremely important.

CHAPTER 5

Case Study: Spatial Architecture Scheduling

5.1 INTRODUCTION

Our third case study is more complex and detailed than the previous two, and pursues a new domain with much broader goals. In this chapter, we will show how MILP can be applied to solve the scheduling problem for a certain class of computer architecture. Similar to the previous case study, we show how a set of constraints can be built upon to model related problems, and we begin this chapter by describing the targeted domain.

Hardware specialization has emerged as an important way to sustain performance improvements of microprocessors to address transistor energy efficiency challenges and general purpose processing's inefficiencies [14, 58, 88]. The fundamental insight of many specialization techniques is to "map" large regions of computation to the hardware, breaking away from instruction-by-instruction pipelined execution and instead adopting a *spatial architecture* paradigm. Specifically, spatial architectures expose hardware resources, like functional units, interconnection network, or storage to the compiler. Pioneering examples include RAW [175], Wavescalar [161], and TRIPS [34], motivated primarily by performance, and recent energy-focused proposals include Tartan [133], CCA [38], PLUG [47, 113], FlexCore [166], SoftHV [48], MESCAL [94], SPL [177], C-Cores [171], DySER [80, 81], BERET [85], and NPU [59].

A fundamental problem in all spatial architectures is the scheduling of some notion of computation to the ISA-exposed hardware resources. Typically this problem has been solved with heuristic based approaches, like the TRIPS and RAW schedulers [40, 116]. Our approach, which leverages mathematical optimization, belongs to the same class as seminal job scheduling and VLIW scheduling work [64, 174]. A mathematical approach can provide many benefits besides solution quality; we seek to create a solution which allows high developer productivity, provides provable properties on results, and enables true architectural generality.

The approach we outline in this chapter captures the scheduling problem with five intuitive abstractions: i) *placement of computation* on the hardware substrate, ii) *routing of data* on the substrate to reflect and carry out the computation semantics—including interconnection network assignment, network contention, and network path assignment, iii) *managing the timing of events* in the hardware, iv) *managing resource utilization* to orchestrate concurrent usage of hardware, and v) forming the *optimization objectives* to meet the architectural performance goals.

Target Architectures   We apply our approach to three architectures picked to stress our MILP scheduler in various ways. To test the performance deliverable by our general MILP approach, we consider TRIPS because it is a mature architecture with sophisticated specialized schedulers resulting from multi-year efforts [3, 34, 40, 138]. To represent the emerging class of energy-centric specialized spatial architectures, we consider DySER [81]. Finally, to demonstrate the generality of our technique, we consider PLUG [47, 113], which uses a radically different organization and execution model. We will show that a total of 20 constraints specify the general problem, and only 3 TRIPS, 1 DySER, and 10 PLUG constraints are required to handle architecture-specific details.

We now briefly describe the three spatial architectures we consider in detail, and a detailed diagram of all three architectures is in Figure 5.8 (page 87).

The **TRIPS architecture** is organized into 16 tiles, with each tile containing 64 slots, with these slots grouped into sets of eight. The slots from one group are available for mapping one block of code, with different groups used for concurrently executing blocks. The tiles are interconnected using a 2-D mesh network, which implements dimension-ordered routing and provides support for flow-control and network contention. The scheduler must perform computation mapping: it takes a block of instructions (which can be no more than 128 instructions long) and assigns each instruction to one of the 16 tiles and within them, to one of the 8 slots.

The **DySER architecture** consists of *functional units* (FUs) and *switches*, and is integrated into the execution stage of a conventional processor. Each FU is connected to four neighboring switches from where it gets input values and injects outputs. The switches allow datapaths to be dynamically specialized. Using a compiler, applications are profiled to extract the most commonly executed regions, called path-trees, which are then mapped to the DySER array. The role of the scheduler is to map nodes in the path-trees to tiles in the DySER array and to determine switch configurations to assign a path for the data-flow graph edges. There is no hardware support for contention, and some mappings may result in unroutable paths. Hence, the scheduler must ensure the mappings are correct, have low latencies and have high throughput.

The **PLUG architecture** is designed to work as an accelerator for data-structure lookups in network processing. Each PLUG tile consists of a set of SRAM banks, a set of no-buffering routers, and an array of statically scheduled in-order cores. The only memory access allowed by a core is to its local SRAM, which makes all delays statically determinable. Applications are expressed as dataflow graphs with code-snippets (the PLUG literature refers to them as code-blocks) and memory associated with each node of the graph. Execution of programs is data-flow driven by messages sent from tile to tile—the ISA provides a *send* instruction. The scheduler must perform computation mapping and network mapping (dataflow edges → networks). It must ensure there is no contention for any network link, which it can do by scheduling when *send* instructions execute in a code-snippet or adjusting the mapping of graph nodes to tiles. It must also handle flow-control.

In all three architectures, multiple instances of a block, region, or dataflow graph are executing concurrently on the same hardware, resulting in additional contention and flow-control.

5.2 OVERVIEW

We present below the main insights of our approach in using MILP for specifying the scheduling problem for spatial architectures. Subsequently, we distill the formulation into five responsibilities, or subproblems, each corresponding to one architectural primitive of the hardware.

The scheduler for a spatial architecture works at the granularity of "blocks" of code, which could be basic-blocks, hyper-blocks, or other code regions. These blocks, which we represent as directed acyclic graphs (DAGs), consist of computation instructions, control-flow instructions, and memory access instructions, which we refer to as G. The hardware, which is composed of functional units and routers, is referred to as H. For ease of explanation, we explain G as comprised of vertices and edges, while H is comprised of nodes, routers, and links (formal definitions and details follow in Section 5.3). An example scheduling problem is shown in Figure 5.1 on page 80.

Problem Statement

Spatially map a typed computation DAG G to a hardware graph H under the architectural constraints.

Though the above problem statement captures the overall problem, the complexity of the problem is reduced through further subdivisions. To design and implement a general scheduler applicable to many spatial architectures, we observe that five fundamental architectural primitives, each with a corresponding scheduler responsibility, capture the problem as outlined in Table 5.1 (columns 2 and 3). Below we describe the insight connecting the primitives and responsibilities and highlight the mathematical approach. Table 5.1 summarizes this correspondence (in columns 2 and 3), and describes these primitives for three different architectures.

Computation HW organization $\rightarrow$ Placement of computation   The spatial organization of the computational resources, which could have a homogeneous or heterogeneous mix of computational units, requires the scheduler to provide an assignment of individual operations to hardware locations. As part of this responsibility, vertices in G are mapped to nodes in the H graph.

Network HW organization $\rightarrow$ Routing of data   The capabilities and organization of the network dictate how the scheduler must handle the mapping of communication between operations to the hardware substrate, i.e., the scheduler must create a mapping from edges in G to the links represented in H. As shown in the second row of Table 5.1, the network organization consists of the spatial layout of the network, the number of networks, and the network routing algorithm. The flow of data required by the computation block and the placement of operations defines the required communication. Depending on the architecture, the scheduler may have to select a network for each message, or even select the exact path it takes.

#	Architecture feature	Scheduler Responsibility	TRIPS	DySER	PLUG
1	Compute HW organization	Placement of computation	Homogeneous compute units	Heterogeneous compute units	Homogeneous compute units
2	Network HW organization	Routing of data	2D grid, dimension order routing	2D grid, unconstrained routing	2D multi-network grid, dimension-order routing
3	HW timing and synchronization	Manage timing of events	Data-flow execution and dynamic network arbitration	Data-flow execution and conflict-free network assignment, flow control	Hybrid data-flow and in-order execution with static compute and network timing
4	Concurrent HW usage within block	Manage resource utilization	8 slots per compute unit, reg-tile, data-tile	No concurrent usage, dedicated compute units, switches, links	32 slots per compute-unit and multicast communication
	Concurrent HW usage across blocks		Concurrent execution of different blocks	Concurrent usage across blocks with pipelined execution	Pipelined execution across different tiles
5	Performance Goal, Architecturally Mandated	Naturally enforced by MILP constraints	Any assignment Legal	Throughput	Throughput
	Performance Goal, High Effeciency	MILP objective formulation	Throughput and Latency	Latency & Latency Mismatch	Latency

Table 5.1: Relationship between architectural primitives and scheduler responsibilities

Hardware timing/synchronization → Manage timing of events   The scheduler must take into consideration the timing properties of computation and network together with architectural restrictions, as shown in the third row of Table 5.1. In some architectures, the scheduler cannot determine the exact timing of events because it is affected by dynamic factors (e.g. memory latency through the caching hierarchy). For all architectures, the scheduler must have at least a partial view of timing of individual operations and individual messages to be able to minimize the latency of the computation block. In some architectures, the scheduler must exert extensive fine-grained control over timing to achieve static synchronization of certain events.

Concurrent hardware resource usage → Managing Utilization   Central to the difficulties of the scheduling problem is the concurrent usage of hardware resources by multiple vertices/edges in G of one node/link in H. We formalize this concurrent usage with a notion of *utilization*, which represents the amount of work a single hardware resource performs. Such concurrent usage (and hence > 1 *utilization*) can occur *within* a DAG and across concurrently executing DAGs. Overall, the scheduler must be aware of resource limits in H and which resources can be shared as shown in Table 5.1, row 4. For example, in TRIPS, within a single DAG, 8 instruction-slots share a single ALU (node in H), and across concurrent DAGs, 64 slots share a single ALU in TRIPS. In both cases, this node-sharing leads to contention on the links as well.

Performance goal $\rightarrow$ Formulate MILP objective   The performance goals of an architecture generally fall into two categories: those which are enforced by certain architectural limitations or abilities, and those which can be influenced by the schedule. For instance, both PLUG and DySER are throughput engines that try perform one computation per cycle, and *any legal* schedule will naturally enforce this behavior. For this type of performance goal, the scheduler relies on the MILP constraints already present in the model. On the other hand, the scheduler generally has control over multiple quantities which can improve the performance. This often means deciding between the conflicting goals of minimizing the latency of individual blocks and managing the utilization among the available hardware resources to avoid creating bottlenecks, which it manages by prioritizing optimization quantities.

5.3   FORMULATION: PARAMETERS AND DECISION VARIABLES

This section presents our general MILP formulation in detail. Our formal notation closely follows our GAMS formulation instead of the more conventional notation often used for graphs in literature. We represent the computation graph as a set of vertices V, and a set of edges E. The computation DAG, represented by the adjacency matrix $G(V \cup E, V \cup E)$, explicitly represents edges as the connections between vertices. For example, for some $v \in V$ and $e \in E$, $G(v, e) = 1$ means that edge e is an output edge from vertex v. Likewise, $G(e, v) = 1$ signifies that e is an input to vertex v. For convenience, lowercase letters represents elements of the corresponding uppercase letters' set.

We similarly represent the hardware graph as a set of hardware computational resource nodes N, a set of routers R which serve as intermediate points in the routing network, and a set of L unidirectional links which connect the routers and resource nodes. The graph which describes the network organization is given by the adjacency matrix $H(N \cup R \cup L, N \cup R \cup L)$. To clarify, for some $l \in L$ and $n \in N$, if the parameter $H(l, n)$ was 0, link l would not be an input of node n. Hardware graphs are allowed to take any shape, and typically do contain cycles. Terms vertex/edge refer to members in G, and node/link to members in H.

Some of the vertices and nodes represent not only computation, but also inputs and outputs. To accommodate this, vertices and nodes are "typed" by the operations they can perform, which also enables the support of general heterogeneity in the architecture. For the treatment here, we abstract the details of the "types" into a compatibility matrix $C(V, N)$, indicating whether a particular vertex is compatible with a particular node. When equations depend on specific types of vertices, we will refer this set as V_{type}.

Figure 5.1 shows an example G graph, representing the computation $z = (x + y)^2$, and an H graph corresponding to a simplified version of the DySER architecture. Here, triangles

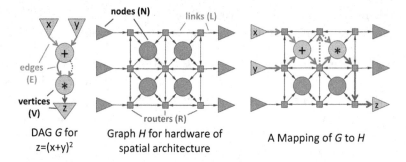

DAG G for $z=(x+y)^2$

Graph H for hardware of spatial architecture

A Mapping of G to H

Figure 5.1: Example of computation G mapped to hardware H.

represent input/output nodes and vertices, and circles represent computation nodes and vertices. Squares represent elements of R, which are routers composing the communication network. Elements of E are shown as unidirectional arrows in the computation DAG, and elements of L as bidirectional arrows in H representing two unidirectional links in either direction.

The scheduler's job is to use the description of the typed computation DAG and hardware graph to find a mapping from computation vertices to computation resource nodes and determine the hardware paths along which individual edges flow. Figure 5.1 also shows a correct mapping of the computation graph to the hardware graph. This mapping is defined by a series of constraints and variables described in the remainder of this section, and these variables and scheduler inputs are summarized in Table 5.3.

5.4   FORMULATION: CONSTRAINTS

We now describe the MILP constraints which pertain to each scheduler responsibility, then show a diagram capturing this responsibility pictorially for our running example in Figure 5.1. Although the MILP solver solves all constraints simultaneously, for exposition we show partial solutions developed at the end of each responsibility.

RESPONSIBILITY 1: PLACEMENT OF COMPUTATION

The first responsibility of the scheduler is to map vertices from the computation DAG to nodes from the hardware graph. Formally, the scheduler must compute a mapping from V to N, which we represent with the matrix of binary variables $M_{vn}(V, N)$. If $M_{vn}(v, n) = 1$ it means that vertex v is mapped to node n, $M_{vn}(v, n) = 0$ it means that v is not mapped to n. Each vertex $v \in V$ must be mapped to exactly one compatible hardware node $n \in N$ in accordance with $C(v, n)$. The mapping for incompatible nodes must also be disallowed. This gives us:

$$\forall v \ \Sigma_{n|C(v,n)=1} M_{vn}(v, n) = 1 \tag{5.1}$$
$$\forall v, n | C(v, n) = 0, \ \ M_{vn}(v, n) = 0 \tag{5.2}$$

Input Parameters: Computation DAG Description (G)	
V	Set of computation vertices.
E	Set of Edges representing data flow of vertices.
$G(V \cup E, V \cup E)$	The computation DAG.
$\Delta(E)$	Delay between vertex activation and edge activation.
$\Delta(V)$	Duration of vertex.
$\Gamma(E)$ (PLUG)	Delay between vertex activation and edge reception.
B_e	Set of bundles which can be overlapped in network.
B_v (PLUG only)	Set of mutually exclusive vertex bundles.
$B(E \cup V, B_e \cup B_v)$	Parameter for edge/vertex bundle membership.
P (TRIPS only)	Set of control flow paths the computation can take.
$A_v(P, V),$ $A_e(P, E)$ (TRIPS)	Activation matrices defining which vertices and edges get activated by given path.

Input Parameters: Hardware Graph Description (H)	
N	Set of hardware resource Nodes.
R	Routers which form the network.
L	Set of unidirectional point-to-point hardware Links.
$H(N \cup R \cup L,$ $N \cup R \cup L)$	Directed graph describing the Hardware.
$I(L, L)$	Link pairs incompatible with Dim. Order Routing.

Inputs: Relationship between G/H	
$C(V, N)$	Vertex-Node compatibility matrix
MAX_N, MAX_L	Maximum degree of mapping for nodes and links.

Variables: Final Outputs	
$M_{vn}(V, N)$	Mapping of computation vertices to hardware nodes.
$M_{el}(E, L)$	Mapping of edges to paths of hardware links.
$M_{bl}(B_e, L)$	Mapping of edge bundles to links.
$M_{bn}(B_v, N)$ (PLUG only)	Mapping of vertex bundles to nodes.
$\delta(E)$ (PLUG)	Padding cycles before message sent.
$\gamma(E)$ (PLUG)	Padding cycles before message received.

Variables: Intermediates	
$O(L)$	The order a link is traversed in.
$U(L \cup N)$	Utilization of links and nodes.
$U_p(P)$ (TRIPS)	Max utilization for each path P.
$T(V)$	Time when a vertex is activated.
$X(E)$	Extra cycles message is buffered.
$\lambda(b, e)$ (PLUG)	Cycle when e is activated for bundle b.
LAT	Total latency for scheduled computation.
SVC	Service interval for computation.
MIS	Largest latency mismatch.

Table 5.2: Summary of formal notation used

Ar

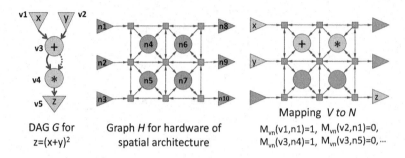

<table>
<tr><td>DAG G for</td><td>Graph H for hardware of</td><td>Mapping V to N</td></tr>
<tr><td>z=(x+y)²</td><td>spatial architecture</td><td></td></tr>
</table>

DAG *G* for
z=(x+y)^2 Graph *H* for hardware of
spatial architecture   Mapping *V to N*
$M_{vn}(v1,n1)=1$, $M_{vn}(v2,n1)=0$,
$M_{vn}(v3,n4)=1$, $M_{vn}(v3,n5)=0$, ...

Figure 5.2: Placement of computation.

RESPONSIBILITY 2: ROUTING OF DATA

The second responsibility of the scheduler is to map the required flow of data to the communication paths in the hardware. We use a matrix of binary variables $M_{el}(E, L)$ to encode the mapping of edges to links. Each edge e must be mapped to a sequence of one or more links l. This sequence must start from and end at the correct hardware nodes. We constrain the mappings such that if a vertex v is mapped to a node n, every edge e leaving from v must be mapped to one link leaving from n. Similarly, every edge arriving to v must be mapped to a link arriving to n.

$$\forall v, e, n | G(v, e), \Sigma_{l|H(n,l)}, M_{el}(e, l) = M_{vn}(v, n) \qquad (5.3)$$
$$\forall v, e, n | G(e, v), \Sigma_{l|H(l,n)}, M_{el}(e, l) = M_{vn}(v, n) \qquad (5.4)$$

In addition, the scheduler must ensure that each edge is mapped to a *contiguous* path of links. We achieve this by enforcing that for each router, either we have no incoming or outgoing links mapped to a given edge, or we have exactly one incoming and exactly one outgoing link mapped to the edge.

$$\forall e \in E, r \in R \quad \Sigma_{l|H(l,r)}, M_{el}(e, l) = \Sigma_{l|H(r,l)} M_{el}(e, l) \qquad (5.5)$$
$$\forall e \in E, r \in R \quad \Sigma_{l|H(l,r)}, M_{el}(e, l) \leq 1 \qquad (5.6)$$

Figure 5.3 shows these constraints applied to the example.

Some architectures require dimension order routing: a message propagating along the X direction may continue on a link along the Y direction, but a message propagating along the Y direction cannot continue on a link along the X direction. To enforce this restriction, we expand the description of the hardware with $I(L, L)$, the set of link pairs that cannot be mapped to the same edge (i.e., an edge cannot be assigned to a path containing any link pair in this set).

$$\forall l, l' | I(l, l'), e \in E, \ M_{el}(e, l) + M_{el}(e, l') \leq 1 \qquad (5.7)$$

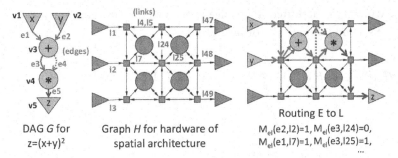

DAG G for $z=(x+y)^2$	Graph H for hardware of spatial architecture	Routing E to L $M_{el}(e2,l2)=1, M_{el}(e3,l24)=0,$ $M_{el}(e1,l7)=1, M_{el}(e3,l25)=1,$ ...

Figure 5.3: Routing of data.

RESPONSIBILITY 3: MANAGE TIMING OF EVENTS

We capture the timing through a set of variables $T(V)$ which represents the time at which a vertex $v \in V$ starts executing. For each edge connecting the vertices v_{src} and v_{dst}, we compute the $T(v_{dst})$ based on $T(v_{src})$. This time is affected by three components. First, we must take into account the $\Delta(E)$, which is the number of clock cycles between the start time of the vertex and when the data is ready. Next is the total routing delay, which is the sum of the number of mapped links between v_{src} and v_{dest}. Since the data carried by all input edges for a vertex might not all arrive at the same time, the variable $X(E)$ describes this mismatch.

$$\forall v_{src}, e, v_{dest} | G(v_{src}, e) \& G(e, v_{dest}),$$
$$T(v_{src}) + \Delta(e) + \Sigma_{l \in L} M_{el}(e, l) + X(e) = T(v_{dest}) \tag{5.8}$$

The equation above cannot fully capture dynamic events like cache misses. Rather than consider all possibilities, the scheduler simply assumes best-case values for unknown latencies (alternatively, these could be attained th̶ ̶ ̶ ̶ ̶ ̶ ̶ ̶ ̶ ̶ ̶ ̶ ̶ ̶ ̶ ̶). Nevertheless, this issue for specialized schedulers as well.

With the constraints thus far, it is possible for the scheduler to overestimate edge latency because the link mapping allows fictitious cycles. As shown by the cycle in the bottom-left quadrant of Figure 5.4, the links in this cycle falsely contribute to the time between input "x" and vertex "+". This does not violate constraint 5.5 because each router involved contains the correct number of incoming/outgoing links.

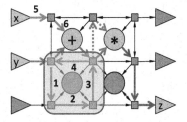

Figure 5.4: Fictitious cycles.

In many architectures, routing constraints (see eq. 5.7) make such loops impossible, but when this is not the case we eliminate cycles through a new constraint. We add a new set of variables $O(L)$, indicating the partial order in which links activated. If an edge is mapped to two connected links, this constraint enforces that the second link must be of later order.

$$\forall l, l', e \in E | H(l, l'), \ O(l) + M_{el}(e, l) + M_{el}(e, l') - 1 \leq O(l') \tag{5.9}$$

Fig ~ ~ ·   ·   ·   ·   ·   ·   ·   ·   · ·

vide.

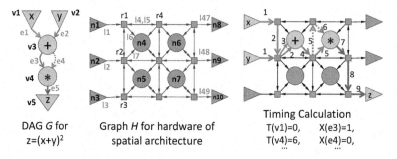

DAG G for   Graph H for hardware of   Timing Calculation
$z=(x+y)^2$ spatial architecture $T(v1)=0, \quad X(e3)=1,$
                                                                       $T(v4)=6, \quad X(e4)=0,$
                                                                       $\dots \qquad \dots$

Figure 5.5: Timing of computation and communication.

RESPONSIBILITY 4: MANAGING UTILIZATION

The utilization of a hardware resource is simply the number of cycles for which it cannot accept a new unit of work (computation or communication) because it is handling work corresponding to another computation. We first discuss the modeling of link utilization $U(L)$, then discuss node utilization $U(N)$.

$$\forall l \in L, \quad U(l) = \Sigma_{e \in E} M_{el}(e, l) \tag{5.10}$$

The equation above models a link's utilization as the sum of its mapped edges and is effective when each edge takes up a resource. On the other hand, some architectures allow for edges to be overlapped, as in the case of multicast, or if it is known that sets of messages are mutually exclusive (will never activate at the same time). This requires us to extend our notion of utilization with the concept of *edge-bundles*, which represent edges that can be mapped to the same link at no cost. The set B_e denotes edge-bundles, and $B(E, B_e)$ defines its relationship to edges. The following three equations ensure the correct correspondence between the mapping of edges to links and bundles to links, and compute the link's work utilization based on the edge-bundles.

$$\forall e, b_e | B(e, b_e), l \in L, \qquad M_{bl}(b_e, l) \geq M_{el}(e, l) \tag{5.11}$$
$$\forall b_e \in B_e, l \in L, \quad \Sigma_{e \in B(e, b_e)} M_{el}(e, l) \geq M_{bl}(b_e, l) \tag{5.12}$$
$$\forall l \in L, \qquad U(l) = \Sigma_{b_e \in B} M_{bl}(b, l) \tag{5.13}$$

To compute the vertices' work utilization, we must additionally consider the amount of time that a vertex fully occupies a node. This time, $\Delta(V)$, is always 1 when the architecture is fully pipelined, but increases when the lack of pipelining limits the use of a node n in subsequent cycles. To compute utilization, we simply sum $\Delta(V)$ over vertices mapped to a node:

$$\forall n \in N \quad U(n) = \Sigma_{v \in V} \Delta(v) M_{vn}(v, l) \tag{5.14}$$

For many spatial architectures we use utilization-limiting constraints such as those below. One application of these constraints are hardware limitations in the number of registers available, instruction slots, etc. Also, they ensure lack of contention with operations or messages from within the same block or other blocks executing concurrently.

$$\forall l \in L, \quad U(l) \leq MAX_L \tag{5.15}$$
$$\forall n \in N, \quad U(n) \leq MAX_N \tag{5.16}$$

As shown in the running DySER example below in Figure 5.6, we limit the utilization of each link $U(l)$ to $MAX_L = 1$. This ensures that only a single message per block traverses the link, allowing the DySER's arbitration-free routers to operate correctly.

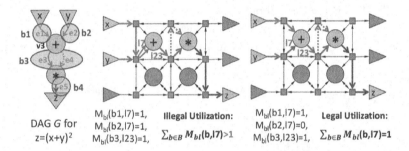

Figure 5.6: Utilization management.

5.5 FORMULATION: OBJECTIVE
RESPONSIBILITY 5: OPTIMIZING PERFORMANCE

The constraints governing the previous sections model the quantities which capture only *individual* components for correctness and performance. However, the final responsibility of the scheduler is to manage the overall correctness while providing performance in the context of the overall system. In practice, this means that the scheduler must balance notions of latency and throughput. Having multiple conflicting targets requires strategic resolution, since there is not necessarily a single solution which optimizes both. The strategy we take is to supply to the scheduler a set of variables to optimize for with their associated priority.

To calculate the critical path latency, we first initialize the input vertices to zero (or some known value) then find the maximum latency of an output vertex LAT. This represents the scheduler's estimate of how long the block would take to complete.

$$\forall v \in V_{in}, \quad T(v) = 0 \tag{5.17}$$
$$\forall v \in V_{out}, \quad T(v) \leq LAT \tag{5.18}$$

To model the throughput aspects, we utilize the concept of the service interval SVC, which is defined as the minimum number of cycles between successive invocations when no data dependencies between invocations exists. We compute SVC by finding the maximum utilization on any resource.

$$\forall n \in N, \quad U(n) \leq SVC \tag{5.19}$$
$$\forall l \in L, \quad U(l) \leq SVC \tag{5.20}$$

For fully pipelined architectures, SVC is naturally forced to 1, so it is not an optimization target. Other notions of throughput are possible, as in the case of DySER, where minimizing the latency mismatch MIS is the throughput objective (see Section 5.6.2).

For our running example, the final solution is shown in Figure 5.7, with critical path latency LAT and the latency mismatch MIS mentioned above, simultaneously optimized by the scheduler

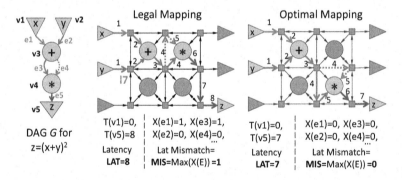

Figure 5.7: Optimizing performance.

5.6 ARCHITECTURE-SPECIFIC MODELING

In this section, we describe how the general formulation presented above is used by three diverse architectures. Figure 5.8 shows schematics and H graphs for the three architectures.

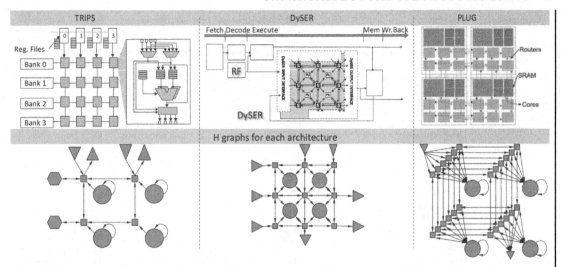

Figure 5.8: Three candidate architectures and corresponding H graphs, considering four tiles for each architecture.

5.6.1   ARCHITECTURE-SPECIFIC DETAILS FOR TRIPS

Computation organization → Placement of computation   Figure 5.8 depicts the graph H we use to describe a 4-tile TRIPS architecture. A tile in TRIPS is comprised of nodes $n \in N$ denoting a functional unit in the tile and $r \in R$ representing its router—the two are connected with one link in either direction. The router also connects to the routers in the neighboring tiles. The functional unit has a self-loop that denotes the bypass of the tile's router to move results into input slots for operations scheduled on the same tile.

Network organization → Routing data   Since messages are dedicated and point-to-point (as opposed to multicast), we use equations modeling each edge as consuming a resource and contributing to the total utilization. The TRIPS routers implement dimension-order routing, i.e., messages first travel along the X axis, then along the Y axis. TRIPS uses the $I(L, L)$ parameter, which disallows the mapping of certain link pairs, to invalidate any paths which are not compatible with dimension-order.

HW timing → Managing timing of events   We can calculate network timing without any additions to the general formulation.

Concurrent HW usage → Utilization   TRIPS allows significant level of concurrent hardware usage which affects both the latency and throughput of blocks. Specifically, the maximum number of vertices per node is $MAX_N = 8$. The utilization on links is used to finally formulate the objective function.

Architecture	Description	MILP Modeling and scheduler responsibility	MILP Constraints
Compute HW Organization	16 tiles, 6 routers per tile	Each tile is 7 nodes in H, 1 in N, and 6 in R	Gen. framework
	4 mem-banks per tile	Handled with utilization	Gen. framework
	32 cores per tile	Handled with utilization	Gen. framework
Network HW Organization	2D nearest neighbor mesh	Node n connected to r; connected to 4 neighbors	Gen. framework
	Dimension order routing	$I(L, L)$ configure for dimension-order routing	Gen. framework
	Multicast messages deliver to every node on path	Map mutlicast message from link to node on path	Constraint 5.25
HW timing	Code-scheduling of network send instructions	Variables for send/receive time	$\Delta(E)$; $\Gamma(E)$
	Send instructions scheduled to avoid network conflicts	Variables for delaying send/receive timing with no-ops	Constraints 5.27a-c, 5.28
Concurrent HW usage	4 mem-banks per time	Manage utilization ($MAX_N = 4$)	Gen. framework
	Dedicated network per message	Manage utilization ($MAX_L = 1$)	Gen. framework
	Mutually exclusive activation of nodes in G	Concept of vertex bundles and utilization refined	Constraints 5.29, 5.30, 5.31
	Code-length limitations (max 32) handled for all code on tile	Manage utilization and combine with vertex bundles	Constraints 5.32, 5.33, 5.34

Table 5.3: Description of MILP model implementation for PLUG

Extensions: For TRIPS, the scheduler must also account for control flow when computing the utilization and ultimately the service interval for throughput. Simple extensions, as explained below, can in general handle control flow for any architecture and could belong in the general MILP formulation as well. Let P be the set of control flow paths that the computation can take through G. Note that $p \in P$ is not actually a path through G, but the subset of its vertices and edges activated in a given execution. Let $A_v(P, V)$ and $A_e(P, E)$ be the activation matrices defining, for each vertex and edge of the computation, whether they are activated when a given path is taken or not. For each path we define the maximum utilization on this path $W_p(P)$. These equations are similar to the original utilization constraints (5.10, 5.14), but also take control flow activation matrices into account.

$$\forall l \in L, p \in P, \quad \Sigma_{e \in E} M_{el}(e, l) A_e(p, e) \leq W_p(p) \tag{5.21}$$
$$\forall n \in N, p \in P, \quad \Sigma_{v \in V} M_{vn}(v, n) \Delta(v) A_v(p, v) \leq W_p(p) \tag{5.22}$$

And an additional constraint for calculating overall service interval:

$$SVC = \Sigma_{p \in P} W_p(p) \tag{5.23}$$

Note that this heuristic provides the same importance to all control-flow paths. With profiling or other analysis, differential weighting can be implemented.

Objective formulation   Empirically, for the TRIPS architecture we found that optimizing for throughput is of higher importance, in most cases, than for latency. Our strategy is to first minimize the SVC, add the lowest value as a constraint, and then optimize for LAT. Below is our solution procedure:

$$min\ SVC\ s.t.\ [5.1\text{–}5.8, 5.10, 5.14\text{–}5.18, 5.21\text{–}5.23]$$
$$min\ LAT\ s.t.\ [5.1\text{–}5.8, 5.10, 5.14\text{–}5.18, 5.21\text{–}5.23] \quad \text{and}\ SVC = SVC_{optimal}$$

5.6.2 ARCHITECTURE-SPECIFIC DETAILS FOR DYSER

Computation organization → Placement of computation   We model DySER with the hardware graph H shown in Figure 5.8; heterogeneity is captured with the $C(V, N)$ compatibility matrix.

Network organization → Routing data   We use bundle-link mapping constraints to model multicast, and Equation 5.9 to prevent fictitious cycles. Since the network has the ability to perform multicast messages and can route multiple edges on the same link, we use the bundle-link mapping constraints. Since there is no ordering constraint on the network, we need to prevent fictitious cycles.

HW timing → Managing timing of events   No additions to the general formulation are required.

Concurrent HW usage → Utilization   Since DySER can only route one message per link, and max one vertex to a node, both MAX_L and MAX_N are set to 1.

Objective Formulation   DySER throughput can be as much as one computation G per cycle, since the functional units themselves are pipelined. However, throughput degradation can occur because of the limited buffering available for messages. The utilization defined in the general framework does not capture this problem because it only measures the usage of functional units and links, not of buffers. Unlike TRIPS, where all operands are buffered as long as needed in named registers, DySER buffers messages at routers and at most one message per edge is buffered at each router. Thus two paths that diverge and then converge, but have different lengths, will also have different amounts of buffering. Combined with backpressure, this can reduce throughput.

Computing the exact throughput achievable by a DySER schedule is difficult, as multiple such pairs of converging paths may exist—even paths that converge after passing through functional units affect throughput. Instead we note that latency mismatches always manifest themselves as extra buffering delays $X(e)$ for some edges, so we model latency mismatch as MIS:

$$\forall e \in E, X(e) \leq MIS \tag{5.24}$$

Empirically, we found that external limitations on the throughput of inputs are greater than that of computation. For this reason, the DySER scheduler first optimizes for latency, adds the latency of the solution as a constraint, then optimizes for throughput by minimizing latency mismatch MIS, as below:

$$min\ LAT\ s.t.\ [5.1\text{–}5.11, 5.12\text{–}5.18, 5.24]$$
$$min\ MIS\ s.t.\ [5.1\text{–}5.11, 5.12\text{–}5.18, 5.24] \quad and\ LAT = LAT_{optimal}$$

5.6.3   ARCHITECTURE-SPECIFIC DETAILS FOR PLUG

The PLUG architecture is radically different from the previous two architectures since all decisions are static. Our formulation is general enough that it works for PLUG with only 10 predominantly simple additional constraints. In the interest of clarity, we summarize the key concepts of the PLUG architecture, corresponding MILP model, and additional equations in Table 5.3. The grayed rows summarize the extensions, and this section's text describes them.

Computation organization $\rightarrow$ Placement of computation   See Table 5.3, row 1. No additional constraints required.

Network organization $\rightarrow$ Routing data   See Table 5.3, row 2.
Additional constraints: Multicast handled with edge-bundles: Let $B_{multi} \subset B_e$ be the subset of edge-bundles that involve multicast edges. The following constraint, which considers links through a router to a node, then enforces that the bundle mapped to the router link must also be mapped to the node's incoming link.

$$\forall b \in B_{multi}, l, r, l', n \,|\, H(l, r) \& H(r, l') \& H(l', n),$$
$$M_{bl}(b, l) \leq M_{bl}(b, l') \tag{5.25}$$

HW timing $\rightarrow$ Managing timing of events   See Table 5.3, row 3.
Additional constraints: We need to handle the timing of send instructions. We use $\Delta(E)$ and the newly introduced $\Gamma(E)$ to respectively indicate the relative cycle number of the corresponding send instruction and use instruction.

Network contention is avoided by code-scheduling the send instructions with NOP padding to create appropriate delays and equalize all delay mismatch. $\delta(E)$ denotes sending delay added, and $\gamma(E)$ denotes receiving delay added. To model the timing for PLUG, we augment Equation 5.8 as follows:

$$\forall v_{src}, e, v_{dst} \,|\, G(v_{src}, e) \& G(e, v_{dst}),$$
$$T(v_{src}) + \Sigma_{l \in L} M_{el}(e, l) + \Delta(e) + \delta(e) = T(v_{dst}) + \Gamma(e) + \gamma(e) \tag{5.26}$$

Because the insertion of no-ops can only change timing in specific ways, we use two constraints to further link $\delta(E)$ and $\gamma(E)$ to $\Delta(E)$ and $\Gamma(E)$. The first ensures that the scheduler never attempts to pad a negative number of NOPs. The second ensures that sending delay $\delta(E)$ is the same for all multicast edges carrying the same message.

To implement these constraints we use the following four sets concerning distinct edges e, e': $SI(e, e')$ is the set of pairs of edges arriving to the same vertex such that $\Gamma(e) < \Gamma(e')$, $LIFO(e, e')$ has for each vertex with both input and output edges the last input edge e and the first output edge e', $SO(e, e')$ has the pairs of output edges with the same source vertex such that $\Delta(e) < \Delta(e')$, and $EQO(e, e')$ has the pairs of output edges leaving the same node concurrently.

$$\forall e, e' | SI(e, e'), \gamma(e) \leq \gamma(e') \tag{5.27a}$$
$$\forall e, e' | LIFO(e, e'), \gamma(e) \leq \delta(e') \tag{5.27b}$$
$$\forall e, e' | SO(e, e'), \delta(e) \leq \delta(e') \tag{5.27c}$$
$$\forall e, e' | EQO(e, e'), \delta(e) = \delta(e') \tag{5.28}$$

Concurrent HW usage $\rightarrow$ Utilization   See Table 5.3, row 4.

Additional constraints: PLUG groups nodes in G into "super-nodes" (logical-page), and programmatically only a single node executes in every super-node. This mutual exclusion behavior is modeled by partitioning V into a set of vertex bundles B_v with $B(V, B_v)$ indicating to which bundle a vertex $v \in V$ belongs. We introduce $M_{bn}(b, n)$ to model the mapping of bundles to nodes, enforced by the following constraints:

$$\forall v, b_v | B(v, b_v), n \in N, \quad M_{bn}(b_v, n) \geq M_{vn}(v, n) \tag{5.29}$$
$$\forall b_v \in B_v, n \in N, \quad \Sigma_{v \in B(v, b_v)} M_{vn}(v, n) \geq M_{bn}(b_v, n) \tag{5.30}$$

We then define the utilization based on the number of vertex bundles mapped to a node. We also instantiate edge bundles b_e for all the set of edges coming from the same vertex bundle and going to the same destination. Since all the edges in such a bundle are *logically* a single message source, the schedule must equalize the receiving times of the message they send. Let $B_{mutex} \subseteq B_e$ be the set of edge-bundles described above. Then we add the following timing constraint:

$$\forall e, e', b_x \in B_{mutex} | B(e, b_x) \& B(e', b_x), \quad \gamma(e) = \gamma(e') \tag{5.31}$$

Additionally, architectural constraints require the total length in instructions of the vertex bundles mapped to the same node to be ≤ 32. This requires defining, for each bundle, the maximum bundle length $\lambda(b_v)$ as a function of the last send message of the vertex. This length can then be constrained to be ≤ 32.

To achieve this, we first define the set $LAST(B_v, B_e)$, which pairs each vertex bundle with its last edge bundle, corresponding to the last send message of the vertex. This enables to define the maximum bundle length $\lambda(b_v)$ as:

$$\forall e, b_e, b_v | LAST(b_v, b_e) \& B(e, b_e), \quad \Delta(e) + \delta(e) \leq \lambda(b_v) \tag{5.32}$$

We finally define $Q(B_v, N)$ as the required number of instructions on node n from vertex bundle b_v and limit it to 32 (the code-snippet length).

$$\forall b_v, n \in N, \quad Q(b_v, n) - 32 * M_{bn}(b_v, n) \geq \lambda(b_v) - 32 \tag{5.33}$$
$$\forall n, \quad \Sigma_{b_v \in B_v} Q(b_v, n) \leq 32 \tag{5.34}$$

Objective Formulation   For PLUG, the smallest service interval is achieved and enforced for any legal schedule, and we optimize solely for latency LAT.

$$min\ LAT\ s.t.\ [5.1\text{--}5.7, 5.11\text{--}5.18, 5.25\text{--}5.34]$$

	TRIPS	DySER	PLUG
Benchmarks	• Same as prior TRIPS scheduler papers [40]. SPEC microbenchmarks and EEBMC • Full SPEC benchmarks can't run to completion on simulator and don't stress scheduler (since blocks are small)	• DySER data-parallel workloads since they produce large blocks and complete code from compiler [80]. • Additional throughput microbenchmark [a]	• PLUG benchmarks from [47]
Native scheduler	• Optimized SPS scheduler [40]	• Specialized greedy algorithm in toolchain & hand scheduled [80]	• Hand scheduled [47]]
Metric	• Total execution cycles for program	• Total execution cycles for program	• Total execution cycles for lookups

Table 5.4: Tools and methodology for quantitative evaluation

[a]DySER "throughput" microbenchmark: This performs the calculation y = x - x 2i in the code-region. Paths diverge at the input node x, into one long path which computes x^{2i} with a series of i multiplies, and along a short path which routes x to the subtraction. This pattern tends to cause latency mismatch because one of these converging paths naturally takes fewer resources.

5.7   MODELING LIMITATIONS

Dynamic Events   Modeling dynamic events, where uncertainty exists in certain problem quantities, is difficult to express in MILP. However, we found that approximating these dynamic events by common-case values was sufficient. We also note that we could extend our model with "stochastic programming" techniques, which solve the same problem for multiple input scenarios. We chose not to explore that for this problem because of the additional model complexity.

Cyclic Computations   This chapter describes how to map computation *DAGs*, which are the typical unit of scheduling. Program loop structures will still occur, just around the unit of a DAG. Some spatial architectures require loops inside the unit of scheduling, but our framework's approach in the calculation of latency would lead to infeasible schedules when considering loops, because each node would have to "come after" the previous one. One simple solution to this problem is to ignore any loop backedges when considering timing, but we have not done a full investigation of an architecture that requires this feature.

Independence of Latency and Utilization   One possible modeling of the spatial scheduling problem is to create binary decision variables both for "where" a computation should go, and "when" it should be activated, which we refer to as "space-time" scheduling. We have taken a slightly different approach in this formulation by assuming that the latency and utilization concerns are mostly independent, and only create decision variables for "where" a computation goes. We rely on the latency being calculable based on the mapping of computation and communication. This is not necessarily true, because with TRIPS, two computations which could both fire on the same tile at the same time will need to be arbitrated. For the purpose of the timing responsibility, the model optimistically assumes that both computations will fire at the same time. In general, our formulation does not take into account the fine-grained interaction of latency and utilization. That said, our approach uses many fewer decision variables than a "space-time" approach, and we can more naturally model utilization constraints.

5.8   EVALUATION

In this section, we describe our implementation of the constraints in and evaluate its performance compared to native specialized schedulers for the three architectures. Our key questions are:

1. Does the MILP model run fast enough to be practical?

2. Are the output solutions produced good? How do they compare against the output of specialized schedulers?

5.8.1   METHODOLOGY

We use the GAMS modeling language to specify our constraints as mixed integer linear programs, and we use the commercial CPLEX solver to obtain the schedules. Our implementation strategy for prioritizing multiple variables follows a standard approach: we define an allowable percentage optimization gap (of between 2% to 10%, depending on the architecture), and optimize for each variable in prioritized succession, finishing the solver when the percent gap is within the specified bounds. After finding the optimal value for each variable, we add a constraint which restricts that variable to be no worse in future iterations.

Figure 5.9 shows our implementation and how we integrated with the compiler/simulator toolchains [3, 47, 80]. For all three architectures, we use their intermediate output converted into our standard directed acyclic graph (DAG) form for G and fed to our GAMS MILP program. We specified H for each architecture. To evaluate our approach, we compare the performance of the final binaries on the architectures varying only the scheduler. Table 5.4 summarizes the methodology and infrastructure used.

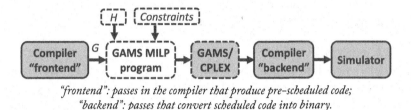

"frontend": passes in the compiler that produce pre-scheduled code;
"backend": passes that convert scheduled code into binary.

Figure 5.9: Implementation of our MILP scheduler. Dotted boxes indicate the new components added.

5.8.2   RESULTS

Practicality   Table 5.5 (page 95) summarizes the mathematical characteristics of the workloads and corresponding scheduling behavior. The three right-hand columns respectively show the number of software nodes to schedule, the amount of single MILP equations created, and the solver time.[1] There is a rough correlation between the workload "size" and scheduling time, but it is still highly variable.

The solver time of the specialized schedulers in comparison is typically in the order of seconds. Although some blocks may take minutes to solve, these times are still tractable, demonstrating the practicality of MILP as a scheduling technique.

Result-1: Our general MILP scheduler runs in tractable time.

Solution Quality vs. Heuristic Schedulers   Figure 5.10 (page 96) shows the performance of our MILP scheduler. It shows the cycle-count reduction for the executed programs as a normalized percentage of the program produced by the specialized compiler (higher is better, negative numbers mean execution time was increased). We discuss these results in terms of each architecture.

Compared to the TRIPS SPS specialized scheduler (a cumulated multi-year effort spanning several publications [34, 40, 138]), our MILP scheduler performs competitively as summarized below.

---

[1]For TRIPS, the per-benchmark number of DAGs can range from 50 to 5000, and the metrics provided are average per DAG. For DySER, #DAGs is 1 to 4 per benchmark, and PLUG is always 1.

Trips μbench	# of nodes	# of eqns	Solve (sec)
ammp_1	17	3744	76
ammp_2	8	1593	11
art_1	22	4547	74
art_2	27	5506	76
art_3	33	7042	20
bzip_1	13	2655	10
equake_1	24	4455	3
gzip_1	23	4480	1
gzip_2	22	4506	111
matrix_1	19	3797	18
parser_1	33	7248	174
transp_GMTI	20	4159	115
vadd	30	7313	315

Trips EEBMC	# of nodes	# of eqns	Solve (sec)
a2time01	11	1914	5
aifftr01	12	2173	25
aifirf01	11	2025	1
basefp01	10	1863	6
bitmnp01	9	1535	3
cacheb01	27	2745	76
candr01	10	1871	8
idctrn01	11	1947	3
iirflt01	11	2080	2
matrix01	11	1426	2
pntrch01	10	1819	8
puwmod01	10	1779	3
rspeed01	10	1816	7
tblook01	10	1818	4
ttsprk1	11	1993	8
cjpeg	12	2280	3
djpeg	12	2277	1
ospf	10	1778	3
pktflow	10	1774	3
routelookup	10	1747	3
bezier01	10	1788	2
dither01	10	3579	4
rotate01	10	1910	5
text01	10	1781	3
autocor00	10	1746	2
conven0	10	1758	4
fbital00	9	1699	3
viterb00	10	1870	5
TRIPS Avg.	**14**	**2832**	**31**

DySER Apps.	# of nodes	# of eqns	Solve (sec)
fft	20	120250	365
mm	32	159231	77
mri-q	19	98615	66
spmv	32	155068	72
stencil	30	153428	74
tpacf	40	211584	368
nnw	25	169197	102
kmeans	40	232399	218
needle	32	181686	183
throughput	9	45138	62
DySER Avg.	**28**	**152660**	**159**

PLUG Apps.	# of nodes	# of eqns	Solve (sec)
Ethernet	18	35603	57
Ethane	11	13905	14
IPv4	12	38741	384
Seattle	16	14531	26
PLUG Avg.	**14**	**23195**	**120**

Table 5.5: Benchmark characteristics and MILP scheduler behavior.

	Compared to SPS		
(a)	Better on 22 of 43 benchmarks	up to 21%	GM +2.9%
(b)	Worse on 18 of 43 benchmarks	within 4.9%	GM -1.9%
		(typically 2%)	
(c)	5.4%, 6.04%, and 13.2% worse on ONLY 3 benchmarks		

Compared to GRST
Consistently better, up to 59% better; GM +30%

Groups (a) and (b) show the MILP scheduler is capturing the architecture/scheduler inter-actions well. The small slowdowns/speedups compared to SPS are due to dynamic events which disrupt the scheduler's view of event timing, making its node/link assignments sub-optimal, typically by only 2%. After detailed analysis, we discovered the reason for the performance gap of group (c) is the lack of information **that could be easily integrated in our model**. First, the SPS scheduler took advantage of information regarding the specific cache banks of loads and stores, which is not available in the modular scheduling interface exposed by the TRIPS compiler. This knowledge would improve the MILP scheduler's performance and would only require changes to the compatibility matrix $C(V, N)$. Second, knowledge of limited resources was available to SPS, allowing it to defer and interact with code-generation to map movement-related instructions. What these results show overall is that our first-principles-based approach is capturing all the architecture behavior in a general fashion and arguably aesthetically cleaner fashion than SPS's indirect heuristics. Our MILP scheduler consistently exceeds by appreciable amounts a previous generation TRIPS scheduler, GRST, that did not model contention [138], as shown by the hatched bars in the figure.

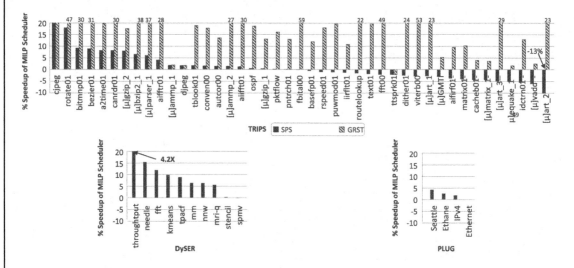

Figure 5.10: Normalized percentage improvement in execution cycles of MILP scheduler compared to special-ized scheduler.

On DySER, the MILP scheduler outperforms the specialized scheduler on all benchmarks, as shown in Figure 5.10, for a 64-unit DySER. Across the benchmarks, the MILP scheduler re-duces *individual* block latencies by 38% on average. When the latency of DySER execution is the bottleneck, especially when there are dependencies between instances of the computation (like the needle benchmark), this leads to significant speedup of up to 15%. We also implemented an extra DySER benchmark, which elucidates the importance of latency mismatch and is described in Table 5.4. The specialized scheduler tries to minimize the extra path length at each step, exacerbating the latency mismatch of the short and long paths in the program. The MILP scheduler, on the other hand, pads the length of the shorter path to reduce latency mismatch, increasing the potential throughput and achieving a **4.2×** improvement over the specialized scheduler. Finally, we also compared to manually scheduled code on a *16-unit* DySER (since hand-scheduling for 64-unit DySER is exceedingly tedious). The MILP scheduler always matched or out-performed it by a small ($< 2\%$) percentage.

The MILP scheduler matches or out-performs the PLUG hand-mapped schedules. It is able to both find schedules that force $SVC = 1$ and provide latency improvements of a few percent. Of particular note is solver time because PLUG's DFGs are more complex. In fact, each DFG represents an *entire* application. The most complex benchmark, IPV4, contains 74 edges (24 more than any others) split between 30 mutually exclusive or multicast groups. Despite these difficulties, it completes in tractable time.

Result-2: Our MILP scheduler outperforms or matches the performance of specialized schedulers.

5.9 RELATED WORK

Many others have used mathematical optimization to address scheduling problems in the past, and we summarize the most related in Table 5.6. In 1950, Wagner describes an MILP model for machine scheduling with dependent tasks [174]. Later, scheduling with MILP is brought to the field of computer architecture, including works like Feautrier's MILP model for modulo scheduling VLIW processors [64], and the multiprocessor scheduling work by Satish et al. [150]. Perhaps the most related work, in terms of the domain, is that of Amarasinghe et al, who formulated an MILP scheduling model for the RAW processor [10]. Though RAW is a spatial architecture, the model differs from ours in that it doesn't perform complex routing of communication, and does not model utilization.

Even though great strides have been made in mathematical models for scheduling, the state-of-the-art approaches for spatial scheduling are heuristic based. The five scheduling abstractions we described are not been modeled directly, and the typically NP-hard (depending on the hardware architecture) spatial architecture scheduling problem is side-stepped. Instead, the focus of architecture-specific schedulers has typically been on developing polynomial-time algorithms that approximate the optimal solution using knowledge about the architecture. Chronologically, this body of work includes the BUG scheduler for VLIW proposed in 1985 [56], UAS scheduler for clustered VLIW [142], synchronous data-flow graph scheduling [17], RAW scheduler [116], CARS VLIW code-generation and scheduler [99], TRIPS scheduler [40, 138], Wavescalar scheduler [130], and CCA scheduler proposed in 2008 [146].

While heuristic-based approaches are popular and effective, they have three problems: i) *poor compiler developer/architect productivity* since new algorithms, heuristics, and implementations are required for each architecture, ii) *lack of insight* on optimality of solution, and iii) *sand-*

Yr	Technique	Comments or differences to our approach
1950	MILP machine sched. [174]	M-Job-DAG to N-resource scheduling. No job communication modeling, or network contention modeling. (missing ii, iv)
1992	MILP for VLIW [64]	Modulo scheduling. Cannot model an interconnection network, spatial resources, or network contention. (missing i, ii, iv)
1997	Inst scheduling [54]	Single-Processor Modulo Scheduling. (missing i, ii, iv)
2001	Process scheduling [57]	M-Job-DAG to N-resource scheduling using dynamic programming. Has no network routing or contention modeling, fixed job delays, and no flexible objective. (missing ii, iv, v)
2002	MILP for RAW [10]	M-Job-DAG to N-resource scheduling. Not generalizable as it does not model network routing or contention, just fixed network delays. (missing ii, iv)
2007	Multiproc. Sched. [110, 150]	M-Job-DAG to N-resource scheduling - No path assignment or contention modeling, just fixed delays. (missing ii, iv)
2008	SMT for PLA [61]	Strict communication and computation requirements: no network contention or path assignment modeling (missing ii, iv)

Table 5.6: Related work – Legend: (i) computation placement; (ii) data routing; (iii) event timing; (iv) work-effort; (v) optimization objective

boxing of heuristics to specific architectures—understanding and using techniques developed for one spatial architecture in another is very hard. Of course, the trade-off in using MILP is that the solution time is significantly longer. However, we were able to create a formulation that runs in a tractable amount of time to be useful for the architectures in question. Furthermore, the declarative approach that MILP enables takes much less development effort than designing an imperative algorithm to perform the same task.

As for other applications of optimization in compilers, uses of MILP in register allocation, code-generation, and optimization-ordering for conventional architectures [98, 144] are unrelated to the primitives of spatial architecture scheduling. Affine loop analysis and resulting instruction scheduling/code-generation for superscalar processors is a popular use of mathematical models [6, 11, 147], and since it falls within the data-dependence analysis role of the compiler, not its scheduler, is not a goal for this work.

5.10   DISCUSSION AND CONCLUSIONS

Scheduling is a fundamental problem for spatial architectures, which are increasingly used to address energy efficiency. This chapter provides a general formulation of spatial scheduling as a MILP problem, demonstrating that all relevant constraints are expressible in MILP. We applied this formulation to three architectures, ran them on a standard MILP solver, and demonstrated such a general scheduler outperforms or matches the respective specialized schedulers.

We conclude with a discussion of some broader extensions and implications of this work, and discuss how our scheduler delivers on its promises of compiler-developer productivity/extensibility, cross-architecture applicability, and insights on optimality.

Formulation Extensibility: In our experience, our model formulation was easily adaptable and extensible for modeling various problem variations or optimizations. For example, we improved upon our TRIPS scheduler's performance by identifying blocks with carried-loop cache dependencies (commonly the most frequently executed), and extended our formulation to only optimize for relevant paths.

Application to Example Architectures: Table 5.7 shows how our framework could be applied to three other systems. For both WaveScalar and RAW, we can attain optimal solutions by refraining from making early decisions, essentially avoiding the drawbacks of multi-stage solutions. For WaveScalar, our scheduler would consider all levels of the network hierarchy at once, using different latencies for links in different networks. For RAW, our scheduler would consider both the partitioning of instructions into streams, and the spatial placement of these instructions simultaneously.

As a more recent example, NPU [59] is a statically scheduled architecture like PLUG, but uses a broadcast network instead of a point-to-point, tiled network. Instead of using the routing equations for communication, the NPU bus is more aptly modeled as a computation type. Timing would be modeled similarly to PLUG, where "no-ops" prevent bus contention, allowing a fully static schedule.

Insights on Optimality: Since our approach provides insights on optimality, it has potentially broader uses as well. For instance, in the context of a dynamic compilation framework, even though the compilation time of seconds is impractical, the MILP scheduler still has significant practical value—it enables developers to easily formulate and evaluate objectives that can guide the implementation of specialized heuristic schedulers.

Revisiting NPU scheduling, we can observe another potential use of MILP models, specifically in designing the hardware itself. For the NPU, the fifo depth of each processing element is expensive in terms of hardware, so we could easily extend the model to calculate the fifo depth as a function of the schedule. One strategy would be to first optimize for performance, then fix the performance and optimize for lowest maximum fifo depth. Doing this across a set of benchmarks would give the best lower-bound fifo depth which does not sacrifice performance.

Finally, while our approach is general, in that we have demonstrated implementations across three disparate architectures and shown extensions to others, an open question remains on "universality": what spatial architecture organization could render our framework ineffective? Overall, our general scheduler can form an important component for future spatial architectures.

Responsibility	RAW	WaveScalar	NPU
Placement	Homogeneous Cores (Tiles)	Homogeneous Processing Elements	8 Processing Elements, 1 Shared Bus
Routing	2D grid, unconstrained routing	Hierarchical Network. First two levels are fully connected, last level grid uses dynamic routing	Responsibility Not Applicable – Broadcast bus used for communication.
Timing	In-order execution inside tile, dataflow between tiles (Secondary list scheduler orders inter-stream events)	Data-flow execution, and dynamic network arbitration; network latency varies by hierarchy level	Fully Static execution. "No-ops" between bus events maintain synchronization.
Utilization	Many instructions per tile. Shared network links	64-Instructions/PE; Shared Network Links	Shared Processing Elements
Objective	Latency & Throughput	Contention & Latency	Latency

Table 5.7: Applicability to other spatial architectures

CHAPTER 6

Case Study: Resource Allocation in Tiled Architectures

6.1 INTRODUCTION

Our final case study looks at a resource allocation problem for "many core" chip multiprocessors. The problem considered in this case study is distinct from the previous ones in two ways. First, while the architectures considered in both chapters are spatially distributed, in this chapter, the hardware itself is responsible for the routing of messages, not the software like in the case of DySER. Second, while our previous case study focused on scheduling operations and communication paths onto given hardware this case study assumes the communication patterns of the spatial components to derive the optimal hardware design.

The specific problem we solve is: given a limited resource budget, determine where memory controllers should be placed on chip and how network links and buffers should be allocated to interconnect processors and memory controllers so as to maximize performance. Though the memory controller placement and network allocation subproblems are individually linear, the combined problem becomes nonlinear, leading to a Mixed Integer Nonlinear Programming model. We then show how to use a reformulation to model the problem as a Mixed Integer Linear Program, greatly improving the degree of optimality guarantees attained, as well as the solution time. Below, we give background on tiled multiprocessors and elaborate on the subproblems.

Tiled Many-Core Architectures   The exponential increases in the number of transistors—commonly known as Moore's Law—has allowed single-chip multiprocessors to grow from multicore processors, with 2-8 cores per processor, to many-core processors, with 64 cores or more. Designers face a combinatorial explosion of alternative ways to allocate a chip's limited resources—e.g., transistors, wires, power, and pins—among cores, caches, on-chip interconnect, and memory controllers. To partially constrain the design space, many-core processors often use a "tiled" architecture, where cores are laid out in a 2D-plane and communicate via an on-chip interconnection network. Tiling restricts many resource allocation problems to deciding in which tile to place a given resource or how many network resources should be allocated to a given interconnect channel. Yet even highly constrained resource allocation problems result in far more configurations than can be evaluated using traditional simulation-based techniques. Feasible solutions can be found using genetic algorithms, but these are time consuming and are not guaranteed to find an optimal solution, motivating our work in applying mathematical optimization.

Memory Controller Placement   Many-core chip multiprocessors require multiple memory controllers to provide the necessary DRAM bandwidth to feed the many cores. In a tiled architecture, a key decision is which tiles should have a memory controller (since the number of memory controllers is typically much smaller than the number of cores). This decision can significantly impact performance because memory controller placement affects both latency and bandwidth [5]. Placement affects the average number of on-chip network hops between a core and memory controller, and thus affects best-case latency. Placement also affects contention for network channels and routers, and thus affects both bandwidth and latency (due to network congestion).

Heterogeneous On-Chip Network Allocation   Tiled architectures can most simply be connected via a homogeneous network, such as a 2D mesh or torus where each router and channel are allocated identical resources. However, recent research has shown that such designs are sub-optimal, since routers near the center of the mesh handle more traffic than those along the edges [132]. Heterogeneous networks allocate network resources—buffers, virtual channels, etc.—differentially, favoring the routers and links that will handle higher traffic loads.

Combined Problem   The memory controller placement problem and the heterogeneous network allocation problem can and have been solved independently. But the two problems are inter-related, since memory controller placement affects the network traffic pattern, and thus the optimal network allocation. This interdependence makes the combined problem non-linear, which has the effect that we are not guaranteed to find an optimal solution. However, it is possible to establish an upper bound on the optimal solution and we show that our results are within 13% of this upper bound. Later, we show how linearizing the constraints through a reformulation technique can fully prove the optimality of the solution.

6.2   OVERVIEW

The problems that we explore are based on the work of Abts, et al. [5] and Mishra, et al. [132]. As in these works, we assume a simplified tiled many-core architecture where cores are laid out on a 2D-plane and connected via an on-chip 2D-switched network that uses a deterministic, dimension-order routing algorithm that ensures that all messages from node A to node B travel the same path. Our modeled architecture approximates systems such as the Tilera TILE-Gx8072 processor [2]. Figure 6.1 shows a high level diagram of a tiled architecture.

Memory Controller Placement   While current tiled architectures typically place memory controllers along the periphery, Abts et al. argue that future designs should distribute them amongst the tiles to reduce average latency and eliminate network hot-spots. Such a design co-locates memory controllers with their assigned cores, so the placement problem simply involves determining which cores are assigned memory controllers, as illustrated in Figure 6.1. However, for n

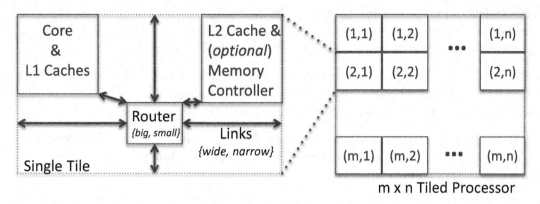

Figure 6.1: Tiled processor with m by n tiles (cores).

cores and m memory controllers, there are $\binom{n}{m}$ possible ways of placing the memory controllers. Thus a 64-core, 16-port design has approximately 4.9×10^{14} possible ways to place the memory controllers, which is well beyond what can be evaluated with typical simulation models. Our approach is to apply integer linear programming to determine memory controller placements that minimize network hot-spots.

We acknowledge that, as part of physical design, there is a second problem where placing tiles with multiple sizes and shapes (because of the added memory controller) could make floor-planning more difficult. We do not address this concern here, and our approach is similar to how others have abstracted this problem [5].

Heterogeneous On-Chip Network Allocation   Current tiled architectures use homogeneous interconnection networks, where each router and network link are provisioned identically, regardless of network traffic. Mishra et al. propose allocating resources to links and routers according to the load they will observe. Specifically, their heterogeneous mesh network is composed of two types of links—*wide* and *narrow*—and two types of routers—*big* and *small*, as illustrated in Figure 6.1.

An 8×8 mesh network requires 64 routers. Assuming 16 big and 48 small routers, there are $\binom{64}{48} \approx 4.89 \times 10^{14}$ possible ways in which these routers can be placed in the network. If the assumption that routers can only be *big* and *small* is dropped, the solution space explodes further. Our approach is to apply integer linear programming to place big routers and wide links in a way that minimizes network contention.

Combined Problem   Since memory controller placement affects network traffic patterns and heterogeneous network allocation affects how much traffic a given router or link can handle before becoming a bottleneck, solving the two problems together should result in a much better overall solution. However, the interactions between the two subproblems results in a non-linear constraint, which requires finding a solution to a mixed integer, non-linear program. While such a problem is incomputable in general [109] (no algorithm exists that can compute a solution), we show that our nonlinear model produces solutions that are within 13% of the optimal solution, and a linear reformulation of the program can prove this solution's optimality.

Problem Statement   Our study focuses on improving throughput in a tiled architecture. Thus we focus on minimizing network contention that can limit throughput and increase latency. Thus we seek to minimize the worst case load on any given link in the interconnect.

> **Determine the placement of memory controllers, router buffers, and link widths which minimizes the worst case, over all links, ratio of per-link traffic to link resources.**

We describe the modeling of the integer linear program by first describing the system abstractly, then writing logical constraints, and where required, linearizing these constraints to match integer linear programming theory. The goal of the next three sections is to describe, from a modeler's perspective, how to formulate the model in terms of the fundamental decision variables, describe the formulation of the constraints by linearizing logical constraints, and finally how to reason about and write the objective.

6.3 FORMULATION: PARAMETERS AND DECISION VARIABLES

The decision variables determine how resources are allocated in our tiled architecture, namely the placement of memory controllers and the assignment of network resources in the heterogeneous network. The system parameters describe the solution space, specifically how the interconnection network routes packets through the network and how much load each memory request places on each router and link.

Decision Variables   Our problem is to find the best, legal placement of memory controllers, router buffers, and link widths to the tiles in our tiled architecture. We assume that there are $N = m \times n$ tiles, organized as m rows of n tiles. Each tile is represented by an ordinal position (x, y), where $1 \leq x \leq m$ and $1 \leq y \leq n$. Each tile has a 2D router with links (x, y, d), where $d \in \{north, east, west, south\}$ (as well as local links to the core and L2 cache bank). In a mesh topology links on the edges of the tiled array are not used, while in a torus the edge links (logically) wrap around to the other side. To hide these topology differences, we consider the set L of valid links $l \in L$ in determining our decision variables.

Memory controller placement is represented by the binary decision variables MC_{xy}, where $MC_{xy} = 1$ implies that tile (x, y) has been assigned a memory controller, and is zero otherwise.

We could similarly model Mishra et al.'s heterogeneous router placement using the binary decision variables BR_{xy}, where $BR_{xy} = 1$ indicates that tile (x, y) is assigned a big router, and zero indicates that it is a small router. However, Mishra et al.'s big/small router formulation unnecessarily restricts all input ports in a router to support the same number of virtual channels and input buffers.[1] We can allocate resources more efficiently if we assign virtual channels and buffers to links independently. Thus, we introduce two new sets of decision variables for heterogeneous network allocation. The variable VC_l is the number of virtual channels assigned to link $l \in L$. Similarly, B_l is the number of input buffers assigned to link $l \in L$.

Finally, heterogeneous link assignments are represented by the decision variables W_l, where $l \in L$, where $W_l = 2$ means that link l is a wide link while $W_l = 1$ means it is a narrow link. While we could have used a binary decision variable to represent which links were wide or narrow, this encoding directly represents the fact that wide links are twice as wide as narrow ones which simplifies our objective function.

System and Workload Parameters   To model our system, we need to present, as input, the parameters which describe the system. For this model, the key parameters describe how messages are routed between two nodes in the system and how much traffic a memory request generates. To model network routing, we introduce the set $Path(x_s, y_s, x_d, y_d)$ which contains all the links l that are used when routing messages from a source tile (x_s, y_s) to a destination tile (x_d, y_d). Note that this formulation is sufficient for any deterministic routing policy; different policies will result in different sets for $Path(x_s, y_s, x_d, y_d)$.

We further introduce two parameters L_{req} and L_{resp}, which indicate the average number of network flits required by a memory request and response, respectively. These values in turn depend upon the cache block size, details of the coherence protocol, and ratio of cache fills to writebacks, but we simplify the formulation to these two parameters.

Auxiliary Variables   A key auxiliary variable, termed $LoadOnLink(l)$, represents the traffic that traverses link l. This variable combines the system and workload parameters with the additional assumption that memory requests are uniformly distributed across the L2 cache banks and memory controllers.

$$
\begin{aligned}
LoadOnLink(l) = & \sum_{(x,y,x',y'):l \in Path(x',y',x,y)} MC_{xy} * L_{req} \\
+ & \sum_{(x,y,x',y'):l \in Path(x,y,x',y')} MC_{xy} * L_{resp} \\
& \text{for all } l \in L
\end{aligned}
\tag{6.1}
$$

Table 6.1 summarizes the parameters and variables of the model.

---

[1]The big/small formulation incorrectly implies that there are exactly two router configurations. In fact, the router's cross-bar is different for each combination of wide/narrow input and output links; thus there are more possible cross-bar configurations than routers in the systems we study.

Input Parameters: Architecture Description	
$Path(x_s, y_s, x_d, y_d)$	Set of network links used when sending a message from tile (x_s, y_s) to tile (x_d, y_d).
L_{req}	Load of a request message.
L_{resp}	Load of a response message.
Input Parameters: Constraint Parameters	
N_{MC}	Number of memory controllers.
N_W	Number of links ($narrow + 2 * wide$).
N_{VC}	Number of virtual channels.
N_B	Number of network buffers.
Variables: Final Outputs	
MC_{xy}	Tile (x, y) includes a memory controller.
W_l	Width of link l.
VC_l	Virtual channels assigned to link l.
Variables: Auxiliary	
$LoadOnLink(l)$	Load carried by link l.

Table 6.1: Summary of formal notation used

6.4   FORMULATION: CONSTRAINTS

We describe the model's constraints in two parts. First, we show how to model the basic design resource constraints that a solution must satisfy to be feasible. Second, we discuss some additional quality constraints, that seek to improve the overall quality of the selected design.

Resource Constraints   Any architectural design problem comes with a set of basic resource constraints. Our model focuses on placing a fixed number of memory controllers among the tiles.

$$\sum_{(x,y)} MC_{xy} = N_{MC} \tag{6.2}$$

This constraint simply states that the number of tiles that are allocated a memory controller is equal to the budget N_{MC}. Note that this constraint must be an equality, since the number of memory controllers is visible to the processor's external interface.

Similar constraints can be stated to enforce the number of wide links, virtual channels, and router buffer budgets.

$$\sum_{l \in L} W_l \quad \leq \quad N_W \qquad (6.3)$$

$$\sum_{l \in L} VC_l \quad \leq \quad N_{VC} \qquad (6.4)$$

$$\sum_{l \in L} B_l \quad \leq \quad N_B \qquad (6.5)$$

Note that since these constraints can be inequalities, since the number actually allocated is not exposed to an external interface.

Quality Constraints   Architects must take into account many factors when evaluating the goodness of a design. In our example, placing too many memory controllers too close together may result in either thermal density (i.e., hot spot) problems or complicate the global routing (e.g., from memory controller to pins). To address these concerns, it may be desirable to add additional constraints to spread the memory controllers. For example, the additional constraints

$$MC_{x,y} + MC_{x,y+1} \quad \leq \quad 1 \qquad (6.6)$$
$$MC_{x,y} + MC_{x+1,y} \quad \leq \quad 1 \qquad (6.7)$$

ensure that adjacent tiles, in rows and columns respectively, are not both assigned memory controllers.

Other quality constraints, which we omit for brevity, include minimum and maximum resource allocations. For example, each link requires at least one virtual channel per virtual network. Similarly, timing constraints may limit the maximum number of buffers that a single link can have without increasing the cycle time.

6.5   FORMULATION: OBJECTIVE

In this section, we show two ways of formulating the objective function. First, we show the intuitive strategy which introduces nonlinearities. Second, we reformulate the problem to linearize these constraints.

Initial Formulation   Our overall objective is to improve system throughput by decreasing network contention. We adopt Abts et al.'s maximum channel load [5] as the figure of merit, thus our goal is to minimize the worst-case utilization of any link. Our goal is to simultaneously minimize the utilization with respect to the resources allocated to each link. We define utilization of a resource for a given link as the ratio of the load on the link and the amount of resource allocated. For example, the utilization of virtual channels for link l is given by: $\dfrac{LoadOnLink(l)}{VC_l}$. We can express our utilization goal by introducing the following auxiliary constraints:

$$
\begin{aligned}
W_l * T &\geq LoadOnLink(l) \text{ for all } l \in L & (6.8) \\
VC_l * S &\geq LoadOnLink(l) \text{ for all } l \in L & (6.9) \\
B_l * W &\geq LoadOnLink(l) \text{ for all } l \in L & (6.10)
\end{aligned}
$$

Here T, S, and W represent the maximum utilization for various per link resources. We set our objective function as: minimize $W + S + T$.

The thing to note is that these auxiliary constraints are not linear, since they involve the ratio of decision variables (the auxiliary variable $LoadOnLink(l)$ is a function of the memory controller placement variables MC_{xy}). Thus our complete formulation is an example of a mixed integer nonlinear program (MINLP).

We can simplify the solution of the problem significantly if we are only interested in the memory controller placement problem. In this case, the network allocation variables W_l, VC_l, and B_l become input parameters, rather than decision variables, and thus the auxiliary constraints become linear. In this case we have an integer linear program, which is much easier to solve.

Similarly, if we are only interested in the heterogeneous network allocation problem [132], we can convert the memory controller placement variables MC_{xy} from decision variables to input parameters. This also makes the auxiliary constraints linear, resulting in an integer linear program.

Linearized Formulation   In our formulation, the only nonlinear constraints are of the form: $f(x, y, w) = w - xy \leq 0$. Thus, each nonlinear term involves the product of two variables. Such terms are called bilinear terms and the formulation is referred to as a Bilinear Program (BLP). Moreover, the bilinear terms in our formulation are products of one continuous and one integer variable. The technique described by Gupte et al., and mentioned in Chapter 2 Section 2.3.5, converts such bilinear terms into linear terms [86]. Doing this allows one to use MILP techniques for solving BLPs.

Consider the constraint $w \leq xy$. Assume that x is a continuous variable and y is an integer variable. Further assume that both of these are non-negative and bounded from above, i.e., $0 \leq x \leq a$ and $0 \leq y \leq b$. Note that constraints (6.8), (6.9), and (6.10) in our formulation are of this form. The set of points satisfying this constraint can be represented as: (where $\mathbb{R}_+$ and $\mathbb{Z}_+$ are positive real numbers and positive integers respectively):

$$
\mathcal{P} = \left\{ (x, y, w) \in \mathbb{R}_+ \times \mathbb{Z}_+ \times \mathbb{R} : w \leq xy, \ x \leq a, \ y \leq b \right\} \tag{6.11}
$$

Using y's binary expansion, we get: $y = \sum_{i=1}^{k} 2^{i-1} z_i$ where z_i are 0-1 integer variables and $k = \lfloor \log_2 b \rfloor + 1$. By forcing v_i to equal xz_i we obtain the following set of linear constraints:

$$\mathcal{B} = \Big\{ (x, y, w, z, v) \in \mathbb{R} \times \mathbb{Z} \times \mathbb{R} \times \{0, 1\}^k \times \mathbb{R}^k :$$

$$y = \sum_{i=1}^{k} 2^{i-1} z_i, \ y \leq b, \ w \leq \sum_{i=1}^{k} 2^{i-1} v_i,$$

$$v_i \geq 0, \ v_i \leq a z_i, \ v_i \leq x, \ v_i \geq x + a z_i - a,$$

$$\text{for all } i \in \{1, \cdots, k\} \Big\} \tag{6.12}$$

It can be shown that $\mathcal{P} = \text{Proj}_{x,y,w}(\mathcal{B})$. Here $\text{Proj}_{x,y,w}$ represents the projection operator that maps $(x, y, w, z, v) \in \mathcal{B}$ to (x, y, w). Note that $\mathcal{B}$ does not have any nonlinear term in its representation and hence it is an exact linearization of $\mathcal{P}$.

We used the approach described above for linearizing this formulation. We replaced each constraint of the form $w \leq xy$, specifically (6.8),(6.9), and (6.10), with the constraints used in defining $\mathcal{B}$.

6.6   MODELING LIMITATIONS

The mixed integer non-linear program we created can effectively model the problem, and with some sophistication can be linearized to find an optimal solution. However, we made certain assumptions about the underlying system, which we describe below:

Deterministic Routing   Our formulation assumed that the routers use a deterministic routing algorithm such as dimension-order routing. While many, if not most, current on-chip routers assume deterministic routing, adaptive routing algorithms have been shown to have superior contention avoidance properties. It is not at all obvious whether it is possible to model adaptive routing in a way that can be solved using mathematical optimization. This is an example of how unknown or dynamic information can preclude the use of optimization.

Uniform Traffic   Our formulation assumed memory accesses were uniformly distributed across the memory controllers and the cores in the tiles. While this is likely to be true over large time periods if scheduling of computation onto cores is uniform, over shorter intervals there may be significant imbalances due to hot memory blocks, etc. Capturing the impact of these effects should be possible, but would make the model more complex.

6.7   EVALUATION

Our evaluation focuses on three important questions:

1. Can the intuitive but non-linear MINLP model be solved to optimality?

Parameter	Value
Processors	64
Memory Controllers	16
Router Latency	2 cycle
Inter-router wire latency	1 cycle
Packet Size	1 flit for request, 5 flits for reply
Virtual Networks	1 for requests, 1 for response
Virtual Channels	Varies with design, router

Table 6.2: Tiled architecture baseline parameters

2. Is it feasible to find a provably optimal solution with the linearized formulation?

3. Do the resulting designs actually perform better than previous designs for real programs?

6.7.1 METHODOLOGY

We wrote the combined model in GAMS. This model describes an abstract representation of a tiled architecture, allowing us to solve for a mathematically optimal solution to the formulation above. To evaluate whether this solution will actually improve performance for real applications, we simulate the execution of workloads drawn from the SPEC CPU2006 benchmark suite [1] using the detailed architectural simulator gem5 [25]. Table 6.2 describes the baseline parameters for the tiled architecture and network that are used in both the mathematical and simulation models.

6.7.2 RESULTS

MINLP Formulation We explored the solution space for the combined problem using Baron [164], an NLP solver, since no efficient methods are known for solving non-convex MINLPs [32]. The solver requires an initial solution to begin with, which is critical since it affects the time required for finding a locally optimal solution and its quality with respect to the globally optimal solution. To get as close as possible to the optimal design, we seeded the solver with multiple different initial designs, and experimented with the bounds on different variables. We selected the initial seeds by first solving the simpler memory controller placement and network allocation problems independently using the Gurobi MILP solver [87], and using those assignments as the seeds. Baron was then able to find designs with improved objective function values.

Figure 6.2 illustrates the best design, referred to as opt, that we obtained from this solution process. Figure 6.2(a), illustrates the placement of the memory controllers and the wide and the narrow links across the mesh network. The filled boxes represent the positions of the memory

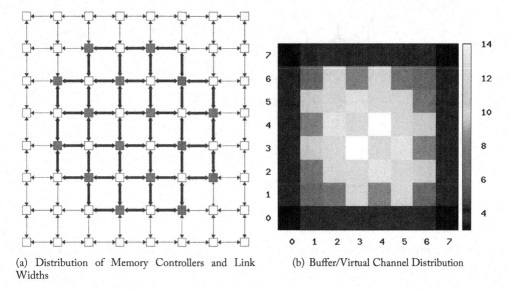

(a) Distribution of Memory Controllers and Link Widths

(b) Buffer/Virtual Channel Distribution

Figure 6.2: Best design For the combined problem.

controllers and the bold lines represent the wide links. Figure 6.2(b) presents a heat map that illustrates the distribution of the virtual channels amongst the routers, where white squares indicate the maximum allocation and dark squares indicate the minimum allocation. It is important to note that neither the memory controller placement nor the allocation of links, virtual channels, and router buffers are the same in the combined problem as they are in the solutions to the respective individual problems.

Our opt design is quite different than the best design evaluated by Mishra, et al. [132], which is illustrated in Figure 6.3. This diagonal design places the memory controllers on the diagonal nodes, leveraging the best solution found by Abts, et al. [5] to the memory controller placement problem. They then place the big routers along the diagonal nodes (6 virtual channels/port) and connect them to their neighbors using wide links. Routers on non-diagonal nodes are small (2 virtual channels/port), and communicate using narrow links, except to neighboring big routers. Note that both the diagonal and the opt designs use the same number of virtual channels, wide and narrow links, and buffers.

Although our NLP solver was not able to guarantee that the opt solution is in fact optimal, it is able to provide a lower bound on the objective value for the optimal solution. This allows us to show that our opt design is within 13% of this lower bound. In comparison, Mishra et al.'s diagonal design is only within 55% of this lower bound on the optimal value.

Result-1: The MINLP model can be solved to within some degree of optimality.

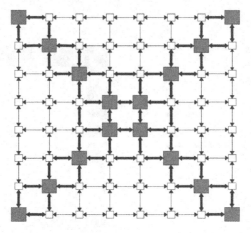

Figure 6.3: Distribution of memory controllers, buffers, and link widths for the diagonal design.

Linearized MILP Formulation   As mentioned in the formulation section, we can reformulate the problem by linearizing the bilinear constraints. We replaced each bilinear constraint with the constraints used in defining set $\mathcal{B}$ in (6.12). Even with many new variables and constraints, this model was vastly more efficient at proving optimality. It took CPLEX [4], fewer than 5 min to prove that the opt design found using the MINLP formulation is in fact optimal for the combined problem.

Result-2: The MILP reformulation can quickly prove the optimality of the solution.

Solution Performance   To determine whether the opt design actually improves performance, we used the architectural simulator gem5 [25] to simulate a detailed architectural model. We used multiprogrammed workloads drawn from the SPEC CPU2006 benchmark suite [1]. For each simulation, we randomly chose eight applications from the suite. Since there are 64 cores in the simulated system, eight copies of each application were simulated. Applications were randomly mapped to the cores. We simulated 44 different combinations and mappings. Each simulation was allowed to run until every processor had executed at least 10,000,000 instructions. We then calculated aggregate IPC (total number of instructions executed by all the processors/total number of clock cycles). Figure 6.4 presents the aggregate IPC of the opt design normalized to that of the diagonal design (i.e., higher is better). We observed an average performance gain of about 10% and no workload performed worse.

Result-3: The optimal solution enables real world performance gains.

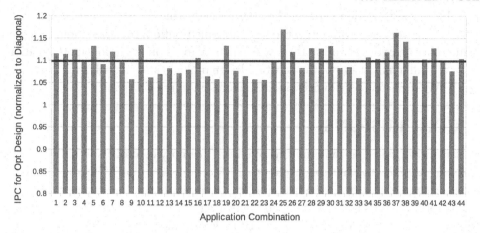

Figure 6.4: Relative IPC for SPEC applications. Each bar is the ratio of aggregate IPC for the opt design and that for the diagonal design.

6.8   RELATED WORK

On-chip Networks   Designing on-chip networks is a fertile area of research. Significant effort has been devoted toward improving the performance and reducing the cost of on-chip networks. Prior works have proposed adaptive routing [122, 123], bufferless NoC [90, 135], and QOS support for NoCs [83]. These works alter the dynamic behavior of the network. Our approach focuses on carrying out the best possible allocation of the resources at the design time. Ben-Itzhak et al. [21] proposed using simulated annealing for designing heterogeneous NoCs. Simulated annealing has scalability problems hence may result in sub-optimal designs. Our work shows that tools for mathematical optimization scale much better.

On-chip Placement   Prior works [12, 176] have focused on figuring out the best mapping for applications and data on to cores and memory controllers at execution time, while we have presented a design time approach. A lot of work for on-chip placement has been done in the SoC domain. These works [95, 184] propose using genetic algorithms for generating solutions.

Xu *et al.* [182] also tackled the problem of placing memory controllers for chip multiprocessors. They solved the problem for a 4 × 4 CMP through exhaustive searched. To find the best placement for the 8 × 8 problem, they exhaustively searched through solutions obtained by stitching solutions obtained for the 4 × 4 problem. This reduces the solution space that needs to be searched, but the idea is not generic. It assumes that the chip can be divided into smaller regions and solutions for the smaller regions can be composed to get optimal solutions for larger regions. This may not hold true in general. Our approach of using mathematical optimization does not rely on any such assumption.

Theory   The network design problem has been widely studied in theoretical computer science [77, 124], particularly with respect to designing distribution, transportation, telecommunication and other types of networks. These works mainly focus on designing approximation algorithms for the different variants of the network design problem and on analyzing their theoretical complexity. We focus on on-chip network design and on-chip placement.

6.9   CONCLUSIONS

In this chapter, we looked at the problems of placing memory controllers in a tiled architecture and allocating resources to a heterogeneous on-chip network. We showed how to formulate these problems and showed that the individual problems result in integer linear programs, but that the combined problem results in a mixed integer non-linear program, which does not guarantee that an optimal solution can be found. We described how we explored the design space for the combined problem using an NLP solver, starting with optimal solutions to the individual problems. We described how the NLP solver can generate an upper bound on the optimal solution, and in our case showed that our best solution was within 13% of this bound. We then described a sophisticated linearization of the formulation, which allowed us to show that our best solution was in fact optimal. Finally, we used detailed architectural simulation to show that our optimal solution improves throughput of multithreaded SPEC CPU2006 workloads by 10% on average.

CHAPTER 7

Conclusions

This book has presented an overview of the modeling technique called mathematical optimization, and has shown how to apply it to solve complex computer architecture problems. We first described optimization in terms of its primitives, and gave a broad overview of related optimization models. Though we did not cover all types of mathematical models in the field of optimization, our broad overview provides the background necessary for interested readers to learn more.

Focusing on one such expressive and efficiently solvable model type, Mixed Integer Linear Programming, we described fundamental modeling techniques relevant to a variety of systems and architecture problems. We then chose four architecture problems to study further, which gave a flavor of the type of problem encountered in the domain. We presented detailed case studies for each, showed how to formulate their problems with MILP, and demonstrated real-world practicality and applicability.

We conclude by summarizing the properties that make problems likely to be amenable to MILP, showing some of MILP's limitations, describing how to decide if MILP is appropriate for a given problem, and finally by describing methods used to improve solve time.

7.1 PROPERTIES OF A MILP-FRIENDLY PROBLEM

Optimization problems encountered in computer architecture, and in systems in general, come in all shapes and sizes. Only a subset of these should be approached with MILP techniques. Broadly speaking, certain problem characteristics make a problem likely to be amenable to this form of mathematical optimization. We elucidate some of these properties below.

Expressible The underlying problem must be expressible as a MILP, meaning that the inputs, constraints, and objectives must be able to be mathematically described within the format of a MILP. An example of inexpressibility is, for a scheduling problem, to maximize the throughput *and* latency. The problem here is that the best throughput and latency might not occur in the same schedule, thus, it is meaningless to "optimize for both". A closely related but expressible model would be to optimize for "the best latency" under a constraint that the actual throughput is within 10% of the best throughput. (The best throughput would be calculated by one run of the model, followed by a second run with a different objective and an additional constraint.)

Tractable Finding the optimal value to a MILP formulation is NP-hard, and unless P=NP, means that solution algorithms are at least exponential in terms of problem size. While the size of the underlying linear program does affect solution times somewhat, it is often the number of integer variables that make the problem hard or even intractable. Therefore, it is wise to limit these to the hundreds to tens of thousands range at this time. Good formulations of the problem are also important, since the enumerative nature of the solution techniques can be improved by generating tight bounds or eliminating many possible alternatives quickly.

Linearity   Simple linear relationships are naturally expressible MILP, and many other conceptual conditions, like logical operations and piecewise-linear functions, can be expressed with auxiliary variables and constraints inside of MILP. Though general nonlinearity can be approximated through certain iterative techniques, a truly linear system will be more efficiently and exactly solvable. We again note that there are other optimization techniques which can directly operate on nonlinear relationships, but this involves complex trade-offs like increased execution time and often lack optimality guarantees. Exploring these techniques is out of the scope of this lecture.

Design Nature   Optimization can and has been used extensively both in design and operational situations. Design problems generally have a "static" nature, in that the problems can be solved off-line, and all input parameters are known at the time of optimization. Resource allocation and compilation are great examples of work which is performed statically, but where the output is used dynamically. This is usually a good fit MILP, as the solution process can be lengthy, but can provide optimality guarantees for problems where the solution is used many times. Operational problems typically require guaranteed bounds on computational time (typically expressed as being in $\mathcal{P}$), and often involve changing data. In this setting, LP and convex programming are thus more applicable, or gradient-based techniques that simply improve the current situation.

7.2   UNDERSTANDING THE LIMITATIONS OF MILP

7.2.1   PROPERTIES OF OPTIMIZATION PROBLEMS UNSUITABLE TO MILP

Below, we describe some of the reasons why MILP might not be an appropriate optimization technique for a given problem, and suggest some solutions. The next subsection examines these issues in the context of two example problems.

Poorly Formulated Problems   Certain problems, even though easily formulated as a MILP, and which are also small enough and have static inputs, can nonetheless still be difficult to solve. This has to do with the nature of the problem itself: if there are many similar quality feasible points to be evaluated, this simply takes time. The intuition behind the mathematics is that the relaxation of certain formulations can be either close to or far from the integer hull of the problem. Refer to Chapter 2 Section 2.4 for details on solution methods. The closer the relaxation, the less work the solver must do in exploring the solution space before finding an optimal integral solution.

Perhaps the most critical component of the formulation is the strength of the root node relaxation. Problems for which the objective value at the root node is close to the actual solution value are much easier to solve. Adding problem-specific cuts (linear inequalities) to the formulation to improve this bound can be very worthwhile.

Problems which have identical solutions due to permutations of a subset of the variables create problems for the branch-and-bound procedure. Identical machines are one such example. Reformulating problems to determine how many machines to use (rather than assigning particular machines to specific tasks) are often effective. Restrictions on an ordering of identical machines also reduces symmetry.

Solution hints: Can extra constraints (cuts) be added to the model? Can the problem be formulated differently, with less symmetry?

Large Problem Size   Every optimization problem has some appropriate bound during which attaining a solution is meaningful and/or useful. A VLSI problem, for instance, maybe be solved in hours or days, as the answer does not need to be immediately returned. For problems in compilers, the range of up to seconds may be appropriate. The more quickly a solution is required, the smaller the problem must be. Genuinely large problems, with millions of equations and/or integer variables, cannot be tractably solved directly with MILP.

Solution hints: Can a problem be effectively solved using a divide-and-conquer technique via multiple MILPs? Can a small instance of the problem be representative of the larger problem?

Nonlinearity   Many optimization problems contain nonlinear relationships between decision variables in the constraints, or can even contain nonlinear objectives. Convex quadratic objectives and constraints are solvable using extensions of the LP technology, but if the nonlinearities cannot be otherwise simplified and linearized, they will make the formulation difficult for MILP solving. This limitation is fundamental; the efficiency of the algorithms used to solve these problems rely on bounds generated using the property of linearity.

Solution hints: Can nonlinearities be approximated by linear or piecewise-linear functions in the domain of interest?

Dynamic Nature   A dynamic optimization problem is one where not all pieces of the problem, generally the input parameters, are known at the time of optimization. This poses a challenge to the ability to formulate the optimization problem at all: how can one write a constraint about an unknown relationship?

Solution hints: Can stochastic programming or robust optimization techniques be applied to account for uncertain future dynamic events?

7.2.2   EXAMPLE PROBLEMS POORLY SUITED TO MILP

In this section, we highlight some optimization problems which are difficult to solve using MILP techniques. Here, we describe the problems, while in the following section we elucidate some of the reasons why they are unmanageable to MILP.

Memory Controller & GPU Warp Scheduling   DRAM controllers in multi-core systems employ out-of-order scheduling techniques to manage parallel accesses. Increasing contention for shared memory causes conflicts for resources, even serializing otherwise parallel operations. The scheduling problem for memory controllers can be formulated as an optimization problem, where the choice of which requests to service in which cycle are the decision variables. Maximizing throughput, minimizing latency, or some combination, would be the natural objective function.

Another interesting scheduling problem occurs in the domain Graphics Processing Units (GPUs). Modern GPUs use logical groups of threads, termed *warps* or *wavefronts*, as the unit of scheduling. Warps are statically assigned to a particular hardware resource, an SM (streaming multiprocessor) or SC (scalar core) (in NVIDIA and AMD terminology respectively). During each cycle of execution, a limited number of threads are scheduled for execution. This per-cycle schedule could be the decision variables for the optimization problem, where minimizing the overall program execution could be the objective function.

Both the Memory Controller Scheduling and GPU Warp Scheduling problems are highly dynamic by nature. What we mean by this is that there is no simple way to formulate the problem such that it can be solved "offline" and employed "online". Each instance of the problem must be solved at the instance it is created: at any cycle of execution, a memory controller must decide which (if any) request to issue, and the GPU warp scheduler must decide which warp to issue. Therefore, the static formulation is significantly less useful. We can use stochastic techniques to model possible events in MILP models, formulating the constraints with an array of particular scenarios, and formulating the objective as an expectation (a weighted sum of individual scenario's objectives).

Even if we were to consider such models, the scheduling problems we considered here become infeasible for another reason: they are orders of magnitude too large. For GPU warp or memory controller scheduling, decision must be made within one to a few cycles, infeasible for any size formulation. Even applying MILP to find an "upper bound" estimate of the schedule might be infeasible, as the length of scheduling trace for the GPU warp or memory controller trace would be likely very large.

Data-center Service-Pattern Allocation   Consider again the data-center allocation problem of Chapter 4. We found ourselves limited to sizes of thousands of services and machines, because we were scheduling each individually. However, if it is indeed true that each machine is identical, and that there are only a certain number of types of services, we can conclude that many machines must have similar service allocations. Here, we define a "service pattern" to be a set of services which will be mapped to a single machine. We could instead formulate the problem as trying to find out which service patterns to use, and how many of each pattern to use, as depicted by Figure 7.1. The constraints are that 1) the total number of services allocated is the total number required, and 2) that no pattern violates the resource requirements of the target machine. The objective is again minimizing the total number of machines required.

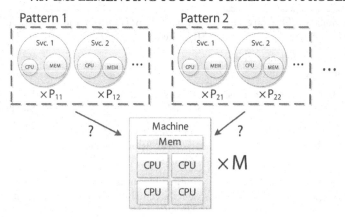

Figure 7.1: Example of the data-center service-pattern allocation problem.

At first glance, this looks to be a very natural optimization problem which is expressible as a MILP and of now tractable size. However, note that we are trying to find the required number of each service pattern which satisfies the total required services of each type. Calculating the total services of each type requires the multiplication of the number of each pattern, and the number of services included in that pattern. This is a quadratic relationship, and serves as an example for how nonlinear constraints can arise in an otherwise linear model.

7.3 IMPLEMENTING YOUR OPTIMIZATION PROBLEMS IN MILP

7.3.1 FIRST STEPS

The first step in getting started with MILP is to ask some basic questions about the nature of the problem. First, it is critical to discern what is driving the model. What are the first-order effects, and what effects can be approximated or treated less precisely? Once this is settled, we suggest looking at the questions outlined in Table 7.1. Answers on the far right indicate that MILP would be a good candidate for the problem. Answers in the middle may require some constraint transformations or optimizations to work properly or efficiently. Any answers on the left hand side indicate significant reformulations or other techniques should be considered. These challenges do arise in practice, and we outline ways to address each of these issues next.

7.3.2 DEALING WITH MILP CHALLENGES

Not Expressible   There are likely two reasons why a problem is not expressible as a MILP. The first is that it is simply difficult to generate linear constraints which describe the system. In the domain of computer architecture, most of the difficulties arise when trying to describe complex logical constraints. For this, we strongly recommend first formulating the problem in terms of

	Reformulation/Other Techniques?	Optimization/Transformations Required?	MILP Suitable
Expressible?	No (cannot define objective function or constraints)		Yes (expressible constraints and objective)
Problem size?	1,000,000+ Eqs. & Variables	10,000s-100,000s of Eqs. & Variables	10s-1000s of Eqs. & Variables
Linearity?	Quadratic or other Non-linear Relationships	Logical, Piece-Wise Linear, or Linear Approximable Relationships	All Linear Relationships
Static vs. Dynamic?	Formulation requires future knowledge, decisions made "on the spot"	Limited Uncertainty: Probable values acceptable or few variables with known distributions	All information known statically

Table 7.1: MILP suitability worksheet

first-order logic, then applying well known techniques to map between logic and integer linear constraints, as demonstrated in Chapter 2. After determining the initial model, more complex and efficient formulations can be explored easily through successive modifications.

The other reason that the problem might not be expressible is that it is difficult to write an objective function which captures many disparate goals. If there is a clear prioritization of goals, then one strategy is to solve the same model multiple times for each objective, adding constraints which enforce previously found objective maxima after each solve. This procedure is referred to as "pre-emptive" goal programming. If, however, a direct comparison of the objective values are required, "non pre-emptive" techniques should be used. Here, the difficulty is in weighting the contribution of multiple objectives to the overall solution. If such a weighting exists, this is likely a preferable solution, as it does not introduce the need for repeated model solution, and the weights often will capture the modelers underlying objective more naturally.

Large Problem Size   Problems which are fundamentally large are very difficult to make amenable to MILP solutions. If MILP is to be applied in such a case, the problem needs to be broken into independent or hierarchical pieces, fixing some decisions at each step. While this no longer can guarantee global optimality, there may still be value in achieving optimality at each step.

Nonlinear Relationships   Some nonlinear relationships, like logical and piecewise-linear constraints covered in Chapter 2, Section 2.3.4, are naturally expressible in MILP through straightforward transformations. More general nonlinear relationships, however, are much more difficult to model directly. There are essentially two possibilities for using optimization on such a problem. The first is to approximate these functions with piecewise-linear functions, which are easily modeled (though more expensive) in MILP. The second approach is to apply a higher order optimization technique like MICQP (mixed integer quadratically constrained programming) or even more generally MINLP (mixed integer nonlinear programming). The trade-offs in using such techniques are many and complex, and are not covered in this book.

Dynamic Problems   Dynamic problems, where some pieces of the problem are unknown at formulation time, can be addressed if some properties of uncertain variables are known. For instance, if we know that a memory latency is either 1, 10, or 100 cycles, with certain probabilities, we can still use linear programming by employing "stochastic programming" techniques. One strategy for solving this problem would be to formulate the constraints for each possible scenario of memory latency, and combine these models into one large MILP. The objective function must quantify the trade-offs between the various scenarios, either by incorporating probabilities and calculating the average objective value, or optimizing for the minimum or maximum objective value across scenarios. These techniques can multiply the size of the model by factors, so care must be taken in selecting representative scenarios.

Poorly Formulated Problems   Formulations with appropriate size, but still performing much too slow, may simply be poorly formulated. Sometimes, reformulations can improve the solution efficiency by orders of magnitude. Though we can only scratch the surface of possible reformulation techniques, we offer some advice below.

When large degrees of symmetry are inherent in the problem, consider adding constraints which disallow certain otherwise feasible possibilities with known equivalents. Recall the data-center allocation problem from Chapter 4, where we added the constraint which enforces that we only "turn on" identical machines in a pre-specified order. This symmetry-breaking constraint does not actually prevent attaining optimal solutions, because there is always an equivalent valid solution for any solution which we disallowed.

We also recommend learning "textbook" reformulation techniques, one of which we describe here as an example. Consider enforcing the logical constraint that any a_i implies b ($\forall_i(a_i \implies b)$). The "aggregated" constraint is:

$$\sum_i a_i \leq M \times b$$

while the disaggregated constraint is:

$$a_i \leq b \text{ for all } i.$$

The aggregated constraint is worse in the sense that relaxation is much further from the integer hull than the disaggregated constraint. In practice, even though the disaggregated version is composed of many constraints, the quality of the formulation invariably leads to faster (and sometimes much faster) solution times.

In some cases, the number of constraints that are needed to describe the convex hull is exponential in the problem data. A good example of this is in traveling salesman problems (Hamiltonian circuits) where "subtour elimination" constraints provide cuts that strengthen the node relaxation problems. In practice, we solve the relaxation and then add in a number of violated constraints of this form—we perform a trade-off between quality of formulation and size of formulation. These tradeoffs should be made empirically with working models.

It is often better to introduce extra variables (that have physical significance) into the model and use them throughout the description of the constraints (as we did in our second example of the introduction, instruction scheduling: page 8). The solver then only has to treat the expression once; the modeler can provide additional information in terms of bounds on these variables, and sometimes branching strategies exploit these "higher level" variables (this can be formalized with priority branching strategies in some solvers).

Adding realistic bounds on as many of the problem variables as possible typically improves performance of the solver. Even if some of these bounds are overly strict, this can help in determining an initial solution to the MILP and thereby checking the logic of the formulation. After solution, some bounds that are artificial, but active, could be relaxed and the problem resolved.

7.3.3   OPTIMIZING AND TUNING MODELS

A frequent problem that arises in solving MILP problems is that the solver determines the model is infeasible. In such cases, it is often difficult to understand the cause of the infeasibility. It is very helpful to "relax" the integrality constraints and see if the underlying linear program is in fact infeasible—modeling systems allow this easily (typically by changing the model type). Fixing a subset of the variables to "realistic" values can sometime help pinpoint the issue in the formulation. In other cases, it is useful to add additional variables to the problem that allow for "slack" in the constraints and replace the objective by a summation of all these slack variables. Solving such an "error minimization" problem often identifies constraints in the model that are incorrectly formulated. These variables can then be removed (or fixed at zero)—sometimes these can be used to determine the benefits of allowing small violations in constraints.

Once a model is solved, it is important to examine the solution values carefully to determine if it passes the "laughability test"—will your colleagues laugh at the given solution? In many cases, hitting the model with extreme (or unlikely) data settings will uncover flaws in the formulation that were not obvious when you formulated the constraints. Additional constraints, and new logic or variables are often needed to flesh out some piece of the formulation so that it correctly describes the physical reality it is attempting to model. This is often time-consuming—in this lecture we have augmented simple observations by more complex simulations—such validation procedures are critical in understanding the limitations of the solution found.

Even a well formulated MILP may still have ample opportunity for improving the speed through other means. For those wishing to further improve their solution times, we offer some advice based on the experience gained in using MILP theory, modeling tools, and solvers.

Relaxing Integrality   Because enforcing integrality constraints can be very costly, due to branching in common MILP solving algorithms, integral variables are generally much more costly than continuous variables. There are at least two circumstances where the integrality conditions on variables can be trivially relaxed. The first case is when the integral variables are themselves dependent only on other integral variables having relationships with only integral coefficients. The integrality of these variables is essentially "free". Explicitly making these variables continuous in the formulation enables solvers to skip these variables as branch targets, potentially improving the solution time significantly.

The other scenario where integral variable constraints could be relaxed is when the variable is itself very large ($>> 1$). Depending on the constraints involved, a solution with an integral variable in the thousands range varying by ± 1 may have very little effect on the optimal objective value. Relaxing the integrality of these variables can, on the other hand, significantly improve the performance of the solver for the same reason as before: these variables do not need to be considered as branch targets, reducing the search space.

Simple is often better. For example, how many segments are really necessary in a piecewise-linear approximation? Do we need to model each instruction, or can some be aggregated together as a group? Is the model a hierarchical model in which fixing some variables then makes the remaining variable choices much easier? Are all the variables measured in the same currency (scaling)? Often simplifications of this form can lead to faster solution times.

Tuning Solvers   As general advice, we recommend always starting with default solver tuning parameters, especially when switching solvers or solver versions. Commercial solvers are fairly intelligent/adaptive. Moreover, adjusting parameters and witnessing speedups on a single problem can be misleading, or even meaningless. Finding speedups on an individual problem can be mere chance rather than true improvements in tweaking the solution strategy. Applying the tuning to a variety of similar problems can help, or automatic tuning can be applied for some solvers.

Adjusting Optimality Requirements   In general, MILP solvers allow the user to specify bounds on required optimality. What this means is that if the solver can prove a certain degree of optimality, then it will return the current incumbent solution. This can lead to huge improvements in performance for the obvious reason that the algorithm can stop early, omitting the exploration of much of the solution space. Making this trade-off is simple, and should be done on a case-by-case basis, depending on the requirements of the problem.

Heuristics for Initial Solutions   Depending on the problem, a MILP solver may spend a significant portion of the solve time without a feasible solution (thereby limiting opportunities for pruning the branch-and-bound search tree). The solver may apply its own internal heuristics to try to find an integer feasible solution, but these are not always effective. Adding domain knowledge to the heuristic can be much more effective, and in these cases it may be a good idea to supply an initial starting solution to the solver. If heuristic procedures for the optimization problem already exist, it is usually very simple to convert these solutions to the modeling language format, and many solvers support this feature. This can significantly reduce the solution time, because the solver can eliminate branches which cannot possibly improve the objective value past the incumbent solution.

7.4   LESSONS LEARNED

In working on these problems and in writing this book, we have learned many things. We share a few of them and our opinions below since they could be beneficial to others.

Expanding scope of problems targetable by MILP   Many interesting design optimization problems in computer architecture benefit from MILP. As the theory behind solving techniques improves, we believe more problems will become practically solvable with MILP and other theories. As a specific example, after completing this work we came across the interesting design problem of interconnect design space in the photonic interconnect work of Kokka et al. [106]. In that work, they use an "ad-hoc" methodology to study various networks and since the optimal network is not known, define an abstract "perfect network" with certain properties to approximate more general designs. On close examination, we realized that this network design problem (and the optimal network problem) can be cast as MILP problem driven by constraints of decibel loss and bandwidth loss, given the network communication patterns.

Spend time upfront to see if MILP seems feasible   Tempering our previous point, we also feel it is important to spend time upfront to understand the fundamental relationships between the variables/system features to determine if the underlying problem is MILP-friendly. For example, we initially felt memory controller scheduling could be cast as a MILP problem. But the nature of the problem, in the sense that it operates on dynamic data, makes such a formulation less useful.

Invest time in tuning the problem and learning the theory   MILP solvers and MILP theory abounds with techniques on reformulation problems to make them "friendlier" for a MILP solver. We encourage readers to invest time in tuning the problem formulation if it seems intractable or too long running on MILP. For example, in the WSAP problem in the second case study, our initial formulation took excessively long in certain cases (more than the 20 min timeout). We then implemented the symmetry breaking constraints, which reduced solver time by three orders of magnitude. We were able to do this transformation because of an understanding of the theory behind solvers. Though this book does *not* provide deep intuition about solvers, our goal is to have encouraged readers and taught readers enough about MILP and its utility and expressiveness so they can invest time in learning about the theory to become expert users of MILP.

If your formulation is looking too complicated, something is wrong   In general if you feel the MILP formulation is too "complicated" and appears non-intuitive, it is very likely you are modeling it the wrong way. Revisiting our spatial scheduling case study, the "complicated" formulation would schedule each operation onto a particular cycle to figure out the cycle-by-cycle resource contention; but the elegant formulation models utilization directly. Unsurprisingly the elegant and intuitive formulation is also solved faster. Look for beauty in your formulation.

We leave the reader with a quote from Paul Dirac on equations which is apt in the context of modeling and fitting models results to experimental data: *"It seems that if one is working from the point of view of getting beauty in one's equations, and if one has really a sound insight, one is on a sure line of progress."*

Bibliography

[1] Standard performance evaluation corporation. http://www.spec.org/results. 110, 112

[2] TILE-Gx8072 Processor. http://www.tilera.com/sites/default/files/productbriefs/TILE-Gx8072_PB041-02.pdf. 102

[3] Trips toolchain, http://www.cs.utexas.edu/~trips/dist/. 76, 93

[4] Ilog cplex 10.1 user's manual, 2006. 112

[5] D. Abts, N. D. Enright Jerger, J. Kim, D. Gibson, and M. H. Lipasti. Achieving Predictable Performance through Better Memory Controller Placement in Many-Core CMPs. In *ISCA '09*. DOI: 10.1145/1555815.1555810. 102, 103, 107, 111

[6] A. V. Aho, M. S. Lam, R. Sethi, and J. D. Ullman. *Compilers: Principles, Techniques, and Tools (2nd Edition)*. Addison-Wesley Longman Publishing Co., Inc., Boston, MA, USA, 2006. 98

[7] R. K. Ahuja, T. L. Magnanti, and J. B. Orlin. *Network Flows: Theory, Algorithms, and Applications*. Prentice-Hall, Englewood Cliffs, NJ, 1993. 1, 23, 24, 28

[8] C. Alippi, W. Fornaciari, L. Pozzi, and M. Sami. A dag-based design approach for reconfigurable vliw processors. In *Proceedings of the conference on Design, automation and test in Europe*, DATE '99, New York, NY, USA, 1999. ACM. DOI: 10.1145/307418.307504. 60

[9] F. Alizadeh and D. Goldfarb. Second-Order Cone Programming. *Mathematical Programming*, 95:3–51, 2003. DOI: 10.1007/s10107-002-0339-5. 14

[10] S. Amarasinghe, D. R. Karger, W. Lee, and V. S. Mirrokni. A theoretical and practical approach to instruction scheduling on spatial architectures. Technical report, MIT, 2002. 97, 98

[11] C. Ancourt and F. Irigoin. Scanning polyhedra with do loops. In *Proceedings of the third ACM SIGPLAN symposium on Principles and practice of parallel programming*, PPOPP '91, pages 39–50, 1991. DOI: 10.1145/109625.109631. 98

[12] M. Awasthi, D. W. Nellans, K. Sudan, R. Balasubramonian, and A. Davis. Handling the Problems and Opportunities Posed by Multiple On-Chip Memory Controllers. In *PACT*, pages 319–330, 2010. DOI: 10.1145/1854273.1854314. 113

[13] M. Azhar, M. Sjalander, H. Ali, A. Vijayashekar, T. Hoang, K. Ansari, and P. Larsson-Edefors. Viterbi accelerator for embedded processor datapaths. In *Application-Specific Systems, Architectures and Processors (ASAP), 2012 IEEE 23rd International Conference on*, pages 133–140, july 2012. DOI: 10.1109/ASAP.2012.24. 4

[14] O. Azizi, A. Mahesri, B. C. Lee, S. J. Patel, and M. Horowitz. Energy-performance trade-offs in processor architecture and circuit design: a marginal cost analysis. In *Proceedings of the 37th annual international symposium on Computer architecture*, ISCA '10, pages 26–36. ACM, 2010. DOI: 10.1145/1816038.1815967. 75

[15] M. Baleani, F. Gennari, Y. Jiang, Y. Patel, R. K. Brayton, and A. Sangiovanni-Vincentelli. Hw/sw partitioning and code generation of embedded control applications on a reconfigurable architecture platform. In *Proceedings of the tenth international symposium on Hardware/software codesign*, CODES '02, pages 151–156, New York, NY, USA, 2002. ACM. DOI: 10.1145/774789.774820. 60

[16] C. Barnhart and E. Johnson. Branch-and-price: Column generation for solving huge integer programs. *Operations Research*, 46(3):316–329, 1998. DOI: 10.1287/opre.46.3.316. 46

[17] S. S. Battacharyya, E. A. Lee, and P. K. Murthy. *Software Synthesis from Dataflow Graphs*. Kluwer Academic Publishers, 1996. DOI: 10.1007/978-1-4613-1389-2. 97

[18] E. Beale and J. Tomlin. Special facilities in a general mathematical programming system for non-convex problems using ordered sets of variables. In J. Lawrence, editor, *Proceedings of the 5th International Conference on Operational Research*, pages 447–454. Tavistock Publishing, London, 1970. 38

[19] E. M. L. Beale and R. E. Small. Mixed Integer Programming by a Branch- and-Bound Technique. In W. A. Kalenich, editor, *IFIP Congress*, pages 450–451. Spartan Press, Washington, D. C, 1965. 2

[20] L. Behjat and A. Chiang. Fast integer linear programming based models for vlsi global routing. In *Circuits and Systems, 2005. ISCAS 2005. IEEE International Symposium on*, pages 6238 – 6243 Vol. 6, may 2005. DOI: 10.1109/ISCAS.2005.1466066. 3

[21] Y. Ben-Itzhak, I. Cidon, and A. Kolodny. Optimizing heterogeneous noc design. SLIP '12, pages 32–39. 113

[22] A. Ben-Tal and A. Nemirovskii. *Lectures on Modern Convex Optimization*. MPS-SIAM Series on Optimization. SIAM, Philadelphia, PA, 2001. 2

[23] J. L. Berral and R. G. amd Jordi Torres. An integer linear programming representation for datacenter power-aware management. Technical report, Universitat Politècnica de Catalunya, 2010. 73

[24] J. L. Berral, R. Gavalda, and J. Torres. Adaptive scheduling on power-aware managed data-centers using machine learning. In *Proceedings of the 2011 IEEE/ACM 12th International Conference on Grid Computing*, GRID '11, pages 66–73, 2011. DOI: 10.1109/Grid.2011.18. 72

[25] N. Binkert, B. Beckmann, G. Black, S. K. Reinhardt, A. Saidi, A. Basu, J. Hestness, D. R. Hower, T. Krishna, S. Sardashti, R. Sen, K. Sewell, M. Shoaib, N. Vaish, M. D. Hill, and D. A. Wood. The gem5 simulator. *SIGARCH Comput. Archit. News*, 39:1–7, Aug. 2011. DOI: 10.1145/2024716.2024718. 110, 112

[26] J. Bisschop and R. Entriken. *AIMMS – The Modeling System*. Paragon Decision Technology, Haarlem, The Netherlands, 1993. 47

[27] J. Bisschop and A. Meeraus. On the Development of a General Algebraic Modeling System in a Strategic Planning Environment. *Mathematical Programming Study*, 20:1–29, 1982. DOI: 10.1007/BFb0121223. 47

[28] R. E. Bixby. A Brief History of Linear and Mixed-Integer Programming Computation. *DocumentA Mathematica*, pages 107–121, 2013. 2

[29] P. Bonami, L. T. Biegler, A. R. Conn, G. Cornuejols, I. E. Grossmann, C. D. Laird, J. Lee, A. Lodi, F. Margot, N. Sawaya, and A. Wachter. An algorithmic framework for convex mixed integer nonlinear programs. *Discrete Optimization*, 5(2):186 – 204, 2008. <ce:title>In Memory of George B. Dantzig</ce:title>. DOI: 10.1016/j.disopt.2006.10.011. 31

[30] S. Bose and S. Sundarrajan. Optimizing migration of virtual machines across data-centers. In *Parallel Processing Workshops, 2009. ICPPW '09. International Conference on*, pages 306–313, Sept. DOI: 10.1109/ICPPW.2009.39. 69, 73

[31] S. Boyd and L. Vandenberghe. *Convex Optimization*, volume 25. Cambridge University Press, 2004. DOI: 10.1017/CBO9780511804441. 48

[32] S. Boyd and L. Vandenberghe. *Convex Optimization*. Cambridge University Press, New York, NY, USA, 2004. DOI: 10.1017/CBO9780511804441. 110

[33] A. Brooke, D. Kendrick, and A. Meeraus. *GAMS: A User's Guide*. The Scientific Press, South San Francisco, California, 1988. 47

[34] D. Burger, S. W. Keckler, K. S. McKinley, M. Dahlin, L. K. John, C. Lin, C. R. Moore, J. Burrill, R. G. McDonald, W. Yoder, and the TRIPS Team. Scaling to the end of silicon with EDGE architectures. *IEEE Computer*, 37(7):44–55, 2004. DOI: 10.1109/MC.2004.65. 75, 76, 94

[35] R. Burkard, M. Dell'Amico, and S. Martello. *Assignment Problems*. Society for Industrial and Applied Mathematics, 2009. 27

[36] K. Chakrabarty. Design of system-on-a-chip test access architectures using integer linear programming. In *VLSI Test Symposium, 2000. Proceedings. 18th IEEE*, pages 127 –134, 2000. DOI: 10.1109/VTEST.2000.843836. 3

[37] K. Chakrabarty. Test scheduling for core-based systems using mixed-integer linear programming. *Computer-Aided Design of Integrated Circuits and Systems, IEEE Transactions on*, 19(10):1163 –1174, oct 2000. DOI: 10.1109/43.875306. 3

[38] N. Clark, M. Kudlur, H. Park, S. Mahlke, and K. Flautner. Application-specific processing on a general-purpose core via transparent instruction set customization. In *Proceedings of the 37th annual IEEE/ACM International Symposium on Microarchitecture*, MICRO 37, pages 30–40, 2004. DOI: 10.1109/MICRO.2004.5. 75

[39] A. Conn, K. Scheinberg, and L. Vicente. *Introduction to derivative-free optimization*. MPS-SIAM Series on Optimization, SIAM, Philadelphia, PA, 2009. DOI: 10.1137/1.9780898718768. 47

[40] K. E. Coons, X. Chen, D. Burger, K. S. McKinley, and S. K. Kushwaha. A spatial path scheduling algorithm for edge architectures. *SIGARCH Comput. Archit. News*, 34(5):129–140, Oct. 2006. DOI: 10.1145/1168919.1168875. 75, 76, 92, 94, 97

[41] R. Cottle, E. Johnson, and R. Wets. George B. Dantzig (1914–2005). *Notices of the AMS*, 54(3):344–362, 2007. 1

[42] R. W. Cottle, J. S. Pang, and R. E. Stone. *The Linear Complementarity Problem*. Academic Press, Boston, MA, 1992. 14

[43] J. Czyzyk, T. Wisniewski, and S. J. Wright. Optimization Case Studies in the NEOS Guide. *SIAM Review*, 41:148–163, 1999. DOI: 10.1137/S0036144598334874. 14

[44] G. Dantzig. On the significance of solving linear programming problems with some integer variables. *Econometrica, Journal of the Econometric Society*, 28(1):30–44, 1960. DOI: 10.2307/1905292. 38

[45] G. B. Dantzig. *Linear Programming and Extensions*. Princeton University Press, Princeton, New Jersey, 1963. 23

[46] Dash Optimization. XPress-Mosel. http://www.fico.com/en/Products/DMTools/xpress-overview/Pages/Xpress-Mosel.aspx, 2009. 47

[47] L. De Carli, Y. Pan, A. Kumar, C. Estan, and K. Sankaralingam. Plug: flexible lookup modules for rapid deployment of new protocols in high-speed routers. In *Proceedings of the ACM SIGCOMM 2009 conference on Data communication*, SIGCOMM '09, pages 207–218, 2009. DOI: 10.1145/1592568.1592593. 75, 76, 92, 93

[48] A. Deb, J. M. Codina, and A. González. Softhv: a hw/sw co-designed processor with horizontal and vertical fusion. In *Proceedings of the 8th ACM International Conference on Computing Frontiers*, CF '11, pages 1:1–1:10, 2011. DOI: 10.1145/2016604.2016606. 75

[49] J. Desrosiers and M. Lübbecke. A primer in column generation. In G. Desaulniers, J. Desrosiers, and M. Solomon, editors, *Column Generation*, volume 3. Springer, 2005. 46

[50] E. W. Dijkstra. A Note on Two Problems in Connexion with Graphs. *Numerische Mathematik*, 1:269–271, 1959. DOI: 10.1007/BF01386390. 26

[51] M. C. DORNEICH and N. V. SAHINIDIS. Global optimization algorithms for chip layout and compaction. *Engineering Optimization*, 25(2):131–154, 1995. DOI: 10.1080/03052159508941259. 3

[52] M. A. Duran and I. E. Grossmann. An outer-approximation algorithm for a class of mixed-integer nonlinear programs. *Mathematical Programming*, 36:307–339, 1986. DOI: 10.1007/BF02592064. 31

[53] L. Eeckhout. *Computer Architecture Performance Evaluation Methods*, volume 5. 2010. 5

[54] A. E. Eichenberger and E. S. Davidson. Efficient formulation for optimal modulo schedulers. In *Proceedings of the ACM SIGPLAN 1997 conference on Programming language design and implementation*, PLDI '97, pages 194–205, 1997. DOI: 10.1145/258915.258933. 98

[55] M. Ekpanyapong, J. Minz, T. Watewai, H.-H. Lee, and S. K. Lim. Profile-guided microarchitectural floor planning for deep submicron processor design. *Computer-Aided Design of Integrated Circuits and Systems, IEEE Transactions on*, 25(7):1289 –1300, july 2006. DOI: 10.1109/TCAD.2005.855971. 3

[56] J. R. Ellis. *Bulldog: a compiler for vliw architectures*. PhD thesis, 1985. 97

[57] D. W. Engels, J. Feldman, D. R. Karger, and M. Ruhl. Parallel processor scheduling with delay constraints. In *Proceedings of the twelfth annual ACM-SIAM symposium on Discrete algorithms*, SODA '01, pages 577–585, 2001. 98

[58] H. Esmaeilzadeh, E. Blem, R. St. Amant, K. Sankaralingam, and D. Burger. Dark silicon and the end of multicore scaling. *SIGARCH Comput. Archit. News*, 39(3):365–376, June 2011. DOI: 10.1145/2024723.2000108. 75

[59] H. Esmaeilzadeh, A. Sampson, L. Ceze, and D. Burger. Neural acceleration for general-purpose approximate programs. In *Proceedings of the 2012 45th Annual IEEE/ACM International Symposium on Microarchitecture*, MICRO '12, pages 449–460, Washington, DC, USA, 2012. IEEE Computer Society. DOI: 10.1109/MICRO.2012.48. 75, 99

[60] F. Facchinei and J. S. Pang. *Finite-Dimensional Variational Inequalities and Complementarity Problems.* Springer-Verlag, New York, New York, 2003. 14

[61] K. Fan, H. h. Park, M. Kudlur, and S. o. Mahlke. Modulo scheduling for highly customized datapaths to increase hardware reusability. In *Proceedings of the 6th annual IEEE/ACM international symposium on Code generation and optimization*, CGO '08, pages 124–133, New York, NY, USA, 2008. ACM. DOI: 10.1145/1356058.1356075. 98

[62] J.-W. Fang, C.-H. Hsu, and Y.-W. Chang. An integer linear programming based routing algorithm for flip-chip design. In *Design Automation Conference, 2007. DAC '07. 44th ACM/IEEE*, pages 606 –611, june 2007. DOI: 10.1109/TCAD.2008.2009151. 3

[63] I. D. Farias, E. Johnson, and G. Nemhauser. Branch-and-cut for combinatorial optimization problems without auxiliary binary variables. *Knowledge Engineering Review*, 16:25–39, 2001. DOI: 10.1017/S0269888901000030. 38

[64] P. Feautrier. Some efficient solutions to the affine scheduling problem. *International Journal of Parallel Programming*, 21:313–347, 1992. DOI: 10.1007/BF01379404. 75, 97, 98

[65] M. C. Ferris and J. S. Pang. Engineering and Economic Applications of Complementarity Problems. *SIAM Review*, 39:669–713, 1997. DOI: 10.1137/S0036144595285963. 3

[66] M. L. Fisher. The Lagrangian Relaxation Method for Solving Integer Programming Problems. *Management Science*, 27:1–18, Dec. 1981. DOI: 10.1287/mnsc.27.1.1. 44

[67] L. Ford and D. R. Fulkerson. A suggested computation for maximal multicommodity network flows. *Management Science*, 5(1):97–101, 1958. DOI: 10.1287/mnsc.5.1.97. 46

[68] L. R. Ford and D. R. Fulkerson. *Flows in Networks.* Princeton University Press, 1962. 23

[69] R. Fourer. On the Evolution of Optimization Modeling Systems. *Documenta Mathematica*, pages 377–388, 2012. 2

[70] R. Fourer, D. M. Gay, and B. W. Kernighan. A Modeling Language for Mathematical Programming. *Management Science*, 36:519–554, 1990. DOI: 10.1287/mnsc.36.5.519. 47

[71] R. Fourer, D. M. Gay, and B. W. Kernighan. *AMPL: A Modeling Language for Mathematical Programming.* Duxbury Press, Pacific Grove, California, 1993. 47

[72] C. Galuzzi and K. Bertels. The instruction-set extension problem: A survey. *ACM Trans. Reconfigurable Technol. Syst.*, 4(2):18:1–18:28, May 2011. DOI: 10.1145/1968502.1968509. 60

[73] C. Galuzzi, E. Panainte, Y. Yankova, K. Bertels, and S. Vassiliadis. Automatic selection of application-specific instruction-set extensions. In *Hardware/Software Codesign and System Synthesis, 2006. CODES+ISSS '06. Proceedings of the 4th International Conference*, pages 160 –165, oct. 2006. DOI: 10.1145/1176254.1176293. 58

[74] GAMS - The solver manuals. http://www.gams.com/dd/docs/solvers/allsolvers.pdf. 41

[75] P. Gilmore and R. Gomory. A Linear Programming Approach to the Cutting-Stock Problem. *Operations Research*, 9(6):849–859, 1961. DOI: 10.1287/opre.9.6.849. 46

[76] F. Glover. Improved linear integer programming formulations of nonlinear integer problems. *Management Science*, 22(4):455–460, 1975. DOI: 10.1287/mnsc.22.4.455. 39

[77] M. X. Goemans, A. V. Goldberg, S. Plotkin, D. B. Shmoys, E. Tardos, and D. P. Williamson. Improved approximation algorithms for network design problems. In *SODA*, pages 223–232, 1994. 114

[78] R. Gomory. OUTLINE OF AN ALGORITHM FOR INTEGER Solutions to Linear Programs. *Bulletin of Mathematical Biophysics*, 64(5):275–278, 1958. DOI: 10.1007/978-3-540-68279-0_4. 2

[79] R. E. Gonzalez. Xtensa: A configurable and extensible processor. *IEEE Micro*, 20(2):60–70, Mar. 2000. DOI: 10.1109/40.848473. 49

[80] V. Govindaraju, C.-H. Ho, T. Nowatzki, J. Chhugani, N. Satish, K. Sankaralingam, and C. Kim. Dyser: Unifying functionality and parallelism specialization for energy efficient computing. *IEEE Micro*, 33(5), 2012. DOI: 10.1109/MM.2012.51. 75, 92, 93

[81] V. Govindaraju, C.-H. Ho, and K. Sankaralingam. Dynamically specialized datapaths for energy efficient computing. In *High Performance Computer Architecture (HPCA), 2011 IEEE 17th International Symposium on*, pages 503–514, 2011. DOI: 10.1109/H-PCA.2011.5749755. 75, 76

[82] M. Grant and S. Boyd. cvx Users' Guide. *Por Clasifcar*, 2(build 711):1–72, 2011. 47, 48

[83] B. Grot, J. Hestness, S. W. Keckler, and O. Mutlu. Kilo-noc: a heterogeneous network-on-chip architecture for scalability and service guarantees. In *ISCA*, pages 401–412, 2011. DOI: 10.1109/MM.2012.18. 113

[84] M. Guevara, B. Lubin, and B. C. Lee. Navigating heterogeneous processors with market mechanisms. In *High Performance Computer Architecture (HPCA2013), 2013 IEEE 19th International Symposium on*, pages 95–106, 2013. DOI: 10.1109/HPCA.2013.6522310. 73

[85] S. Gupta, S. Feng, A. Ansari, S. Mahlke, and D. August. Bundled execution of recurring traces for energy-efficient general purpose processing. In *Proceedings of the 44th Annual IEEE/ACM International Symposium on Microarchitecture*, MICRO-44 '11, pages 12–23, 2011. DOI: 10.1145/2155620.2155623. 75

[86] A. Gupte, S. Ahmed, M. S. Cheon, and S. S. Dey. Solving mixed integer bilinear problems using milp formulations. Technical report, 2012. DOI: 10.1137/110836183. 31, 40, 108

[87] I. Gurobi Optimization. Gurobi optimizer reference manual. http://www.gurobi.com, 2012. 110

[88] N. Hardavellas, M. Ferdman, B. Falsafi, and A. Ailamaki. Toward dark silicon in servers. *IEEE Micro*, 31(4):6–15, 2011. DOI: 10.1109/MM.2011.77. 75

[89] W. E. Hart, C. Laird, J.-P. Watson, and D. L. Woodruff. *Pyomo - Optimization Modeling in Python*. 2012. DOI: 10.1007/978-1-4614-3226-5. 47

[90] M. Hayenga, N. E. Jerger, and M. Lipasti. Scarab: a single cycle adaptive routing and bufferless network. In *MICRO 42*, pages 244–254, 2009. DOI: 10.1145/1669112.1669144. 113

[91] S. Held, B. Korte, D. Rautenbach, and J. Vygen. Combinatorial optimization in vlsi design. *Combinatorial optimization methods and applications*, 31:33–96, 2011. 3

[92] F. L. Hitchcock. The Distribution of a Product from Several Sources to Numerous Facilities. *Journal of Mathematical Physics*, 20:224–230, 1941. 23

[93] A. Hoffmann, H. Meyr, and R. Leupers. *Architecture Exploration for Embedded Processors with Lisa*. Kluwer Academic Publishers, Norwell, MA, USA, 2002. DOI: 10.1007/978-1-4757-4538-2. 49

[94] Z. Huang, S. Malik, N. Moreano, and G. Araujo. The design of dynamically reconfigurable datapath coprocessors. *ACM Trans. Embed. Comput. Syst.*, 3(2):361–384, May 2004. DOI: 10.1145/993396.993403. 75

[95] W.-L. Hung, Y. Xie, N. Vijaykrishnan, C. Addo-Quaye, T. Theocharides, and M. Irwin. Thermal-aware floorplanning using genetic algorithms. In *International Symposium on Quality of Electronic Design, 2005.*, pages 634 – 639. DOI: 10.1109/ISQED.2005.122. 113

[96] R. Jeroslow. Representability in mixed integer programmiing, I: characterization results. *Discrete Applied Mathematics*, 17:223–243, 1987. DOI: 10.1016/0166-218X(87)90026-6. 22

[97] I.-R. Jiang. Generic integer linear programming formulation for 3d ic partitioning. In *SOC Conference, 2009. SOCC 2009. IEEE International*, pages 321 –324, sept. 2009. DOI: 10.1109/SOCCON.2009.5398032. 3

[98] R. Joshi, G. Nelson, and K. Randall. Denali: a goal-directed superoptimizer. In *Proceedings of the ACM SIGPLAN 2002 Conference on Programming language design and implementation*, PLDI '02, pages 304–314, 2002. DOI: 10.1145/512529.512566. 98

[99] K. Kailas, A. Agrawala, and K. Ebcioglu. Cars: A new code generation framework for clustered ilp processors. In *Proceedings of the 7th International Symposium on High-Performance Computer Architecture*, HPCA '01, pages 133–, 2001. DOI: 10.1109/HPCA.2001.903258. 97

[100] J. Kallrath, editor. *Algebraic Modeling Systems: Modeling and Solving Real World Optimization Problems*. Springer Verlag, 2012. DOI: 10.1007/978-3-642-23592-4. 2

[101] L. V. Kantorovich. Mathematical Methods in the Organization and Planning of Production. *Management Science*, 6:366–422, 1960. DOI: 10.1287/mnsc.6.4.366. 23

[102] N. Karmarkar. A New Polynomial Time Algorithm for Linear Programming. *Combinatorica*, 4:373–395, 1984. DOI: 10.1007/BF02579150. 1

[103] A. B. Keha, I. R. de Farias, and G. L. Nemhauser. Models for representing piecewise linear cost functions. *Operations Research Letters*, 32(1):44–48, Jan. 2004. DOI: 10.1016/S0167-6377(03)00059-2. 38

[104] A. B. Keha, I. R. de Farias, and G. L. Nemhauser. A Branch-and-Cut for Nonconvex Algorithm Piecewise Without Binary Variables Linear Optimization. *Operations Research*, 54(5):847–858, 2006. DOI: 10.1287/opre.1060.0277. 36

[105] L. G. Khachiyan. A Polynomial Algorithm for Linear Programming. *Soviet Mathematics Doklady*, 20:191–194, 1979. 1

[106] P. Koka, M. McCracken, H. Schwetman, C.-H. Chen, X. Zheng, R. Ho, K. Raj, and A. Krishnamoorthy. A micro-architectural analysis of switched photonic multi-chip interconnects. In *Computer Architecture (ISCA), 2012 39th Annual International Symposium on*, pages 153–164, 2012. DOI: 10.1109/ISCA.2012.6237014. 124

[107] T. G. Kolda, R. M. Lewis, and V. Torczon. Optimization by Direct Search: New Perspectives on Some Classical and Modern Methods. *SIAM Review*, 45(3):385–482, 2003. DOI: 10.1137/S003614450242889. 3

[108] T. Koopmans. Optimum utilization of the transportation system. *Econometrica: Journal of the Econometric Society*, 17:136–146, 1949. DOI: 10.2307/1907301. 23

[109] M. Koppe. On the Complexity of Nonlinear Mixed-Integer Optimization. In J. Lee and S. Leyffer, editors, *Mixed Integer Nonlinear Programming*, volume 154 of *The IMA Volumes in Mathematics and its Applications*, pages 533–557. Springer New York, 2012. 104

[110] M. Kudlur and S. Mahlke. Orchestrating the execution of stream programs on multicore platforms. In *Proceedings of the 2008 ACM SIGPLAN conference on Programming language design and implementation*, PLDI '08, pages 114–124, 2008. DOI: 10.1145/1375581.1375596. 98

[111] H. W. Kuhn and A. W. Tucker. Nonlinear Programming. In J. Neyman, editor, *Proceedings of the Second Berkeley Symposium on Mathematical Statistics and Probability*, pages 481–492. University of California Press, Berkeley and Los Angeles, California, 1951. 2

[112] C. Kuip. Algebraic languages for mathematical programming. *European Journal of Operational Research*, 67:25–51, 1993. DOI: 10.1016/0377-2217(93)90320-M. 2

[113] A. Kumar, L. De Carli, S. J. Kim, M. de Kruijf, K. Sankaralingam, C. Estan, and S. Jha. Design and implementation of the plug architecture for programmable and efficient network lookups. In *Proceedings of the 19th international conference on Parallel architectures and compilation techniques*, PACT '10, pages 331–342, 2010. DOI: 10.1145/1854273.1854316. 75, 76

[114] F. Larumbe and B. Sansò. Optimal location of data centers and software components in cloud computing network design. In *Cluster, Cloud and Grid Computing (CCGrid), 2012 12th IEEE/ACM International Symposium on*, pages 841–844, May. DOI: 10.1109/CCGrid.2012.124. 62

[115] J.-H. Lee, Y.-C. Hsu, and Y.-L. Lin. A new integer linear programming formulation for the scheduling problem in data path synthesis. In *Computer-Aided Design, 1989. ICCAD-89. Digest of Technical Papers., 1989 IEEE International Conference on*, pages 20 –23, nov 1989. DOI: 10.1109/ICCAD.1989.76896. 3

[116] W. Lee, R. Barua, M. Frank, D. Srikrishna, J. Babb, V. Sarkar, and S. Amarasinghe. Space-time scheduling of instruction-level parallelism on a raw machine. In *Proceedings of the eighth international conference on Architectural support for programming languages and operating systems*, ASPLOS VIII, pages 46–57, 1998. DOI: 10.1145/291069.291018. 75, 97

[117] C. E. Lemke. The dual method of solving the linear programming problem. *Naval Research Logistics Quarterly*, 1(1):36–47, 1954. DOI: 10.1002/nav.3800010107. 1

[118] Y. Lin and L. Schrage. The global solver in the LINDO API. *Optimization Methods and Software*, (4-5):657–668, 2009. DOI: 10.1080/10556780902753221. 31

[119] M. Lobo, L. Vandenberghe, S. Boyd, and H. Lebret. Applications of second-order cone programming. *Linear Algebra and its Applications*, 284:193–228, 1998. DOI: 10.1016/S0024-3795(98)10032-0. 14

[120] R. Lougee-Heimer. The Common Optimization INterface for Operations Research: Promoting open-source software in the operations research community. *IBM Journal of Research and Development*, 47(1):57–66, 2003. DOI: 10.1147/rd.471.0057. 48

[121] B. Lubin, J. O. Kephart, R. Das, and D. C. Parkes. Expressive power-based resource allocation for data centers. In *Proceedings of the 21st international jont conference on Artifical intelligence*, IJCAI'09, pages 1451–1456, San Francisco, CA, USA, 2009. Morgan Kaufmann Publishers Inc. 73

[122] S. Ma, N. Enright Jerger, and Z. Wang. Dbar: an efficient routing algorithm to support multiple concurrent applications in networks-on-chip. In *ISCA-38*, pages 413–424, 2011. DOI: 10.1145/2024723.2000113. 113

[123] S. Ma, N. D. E. Jerger, and Z. Wang. Whole packet forwarding: Efficient design of fully adaptive routing algorithms for networks-on-chip. In *HPCA*, pages 467–478, 2012. DOI: 10.1109/HPCA.2012.6169049. 113

[124] T. L. Magnanti and R. T. Wong. Network Design and Transportation Planning: Models and Algorithms. *Transportation Science*, 18:1–56, 1984. DOI: 10.1287/trsc.18.1.1. 114

[125] H. Markowitz and A. Manne. On the solution of discrete programming problems. *Econometrica: Journal of the Econometric*, 25(1):84–110, 1957. DOI: 10.2307/1907744. 38

[126] J. Mars, L. Tang, R. Hundt, K. Skadron, and M. L. Soffa. Bubble-up: increasing utilization in modern warehouse scale computers via sensible co-locations. In *Proceedings of the 44th Annual IEEE/ACM International Symposium on Microarchitecture*, MICRO-44 '11, pages 248–259, 2011. DOI: 10.1145/2155620.2155650. 63

[127] R. K. Martin. *Large scale linear and integer optimization*. Kluwer Academic Publishers, Boston, MA, 1999. DOI: 10.1007/978-1-4615-4975-8. 39

[128] Maximal Software. MPL. http://www.maximal-usa.com/mpl/what.html, 2009. 47

[129] G. P. McCormick. Computability of Global Solutions to Factorable Nonconvex Programs: Part I - Convex Underestimating Problems. *Mathematical Programming*, 10:147–175, 1976. DOI: 10.1007/BF01580665. 39

[130] M. Mercaldi, S. Swanson, A. Petersen, A. Putnam, A. Schwerin, M. Oskin, and S. J. Eggers. Instruction scheduling for a tiled dataflow architecture. In *Proceedings of the 12th international conference on Architectural support for programming languages and operating systems*, ASPLOS XII, pages 141–150, 2006. DOI: 10.1145/1168857.1168876. 97

[131] R. Misener and C. A. Floudas. GloMIQO: Global Mixed-Integer Quadratic Optimizer. *Journal of Global Optimization*, 2012. DOI: 10.1007/s10898-012-9874-7. 40

[132] A. K. Mishra, N. Vijaykrishnan, and C. R. Das. A case for heterogeneous on-chip inter-connects for cmps. In *Proceedings of the 38th annual international symposium on Computer architecture*, ISCA '11, pages 389–400, 2011. DOI: 10.1145/2024723.2000111. 102, 108, 111

[133] M. Mishra, T. J. Callahan, T. Chelcea, G. Venkataramani, S. C. Goldstein, and M. Budiu. Tartan: evaluating spatial computation for whole program execution. In *Proceedings of the 12th international conference on Architectural support for programming languages and operating systems*, ASPLOS XII, pages 163–174, 2006. DOI: 10.1145/1168857.1168878. 75

[134] G. Mitra and C. Lucas. Tools for reformulating logical forms into zero-one mixed integer programs. *European Journal of Operational Research*, 72:262–276, 1994. DOI: 10.1016/0377-2217(94)90308-5. 34

[135] T. Moscibroda and O. Mutlu. A case for bufferless routing in on-chip networks. In *Proceedings of the 36th annual international symposium on Computer architecture*, ISCA '09, pages 196–207, 2009. DOI: 10.1145/1555815.1555781. 113

[136] B. A. Murtagh and M. A. Saunders. MINOS 5.0 User's Guide. Technical Report SOL 83.20, Systems Optimization Laboratory, Department of Operations Research, Stanford University, Stanford, California, 1983. 2

[137] M. N. Bennani and D. A. Menasce. Resource allocation for autonomic data centers using analytic performance models. In *Proceedings of the Second International Conference on Automatic Computing*, ICAC '05, pages 229–240, Washington, DC, USA, 2005. IEEE Computer Society. DOI: 10.1109/ICAC.2005.50. 72

[138] R. Nagarajan, S. K. Kushwaha, D. Burger, K. S. McKinley, C. Lin, and S. W. Keckler. Static placement, dynamic issue (spdi) scheduling for edge architectures. In *Proceedings of the 13th International Conference on Parallel Architectures and Compilation Techniques*, PACT '04, pages 74–84, 2004. DOI: 10.1109/PACT.2004.26. 76, 94, 96, 97

[139] G. L. Nemhauser and L. A. Wolsey. *Integer and Combinatorial Optimization*. John Wiley & Sons, New York, NY, 1988. 2, 24, 28, 40

[140] R. Niemann and P. Marwedel. Hardware/software partitioning using integer programming. In *European Design and Test Conference, 1996. ED TC 96. Proceedings*, pages 473–479, mar 1996. DOI: 10.1109/EDTC.1996.494343. 4

[141] J. Nocedal and S. J. Wright. *Numerical Optimization*. Springer, New York, 1999. DOI: 10.1007/b98874. 14

[142] E. Özer, S. Banerjia, and T. M. Conte. Unified assign and schedule: a new approach to scheduling for clustered register file microarchitectures. In *Proceedings of the 31st annual ACM/IEEE international symposium on Microarchitecture*, MICRO 31, pages 308–315, 1998. DOI: 10.1109/MICRO.1998.742792. 97

[143] C. Ozturan, G. Dundar, and K. Atasu. An integer linear programming approach for identifying instruction-set extensions. In *Hardware/Software Codesign and System Synthesis, 2005. CODES+ISSS '05. Third IEEE/ACM/IFIP International Conference on*, pages 172 – 177, sept. 2005. 50

[144] J. Palsberg and M. Naik. Ilp-based resource-aware compilation, 2004. 98

[145] P. M. Pardalos and M. G. C. Resende, editors. *Handbook of Applied Optimization*. Oxford University Press, New York, NY, 2002. DOI: 10.1007/978-1-4757-5362-2. 3

[146] H. Park, K. Fan, S. A. Mahlke, T. Oh, H. Kim, and H.-s. Kim. Edge-centric modulo scheduling for coarse-grained reconfigurable architectures. In *Proceedings of the 17th international conference on Parallel architectures and compilation techniques*, PACT '08, pages 166–176, 2008. DOI: 10.1145/1454115.1454140. 97

[147] W. Pugh. The omega test: a fast and practical integer programming algorithm for dependence analysis. In *Supercomputing '91*. DOI: 10.1145/125826.125848. 98

[148] L. M. Rios and N. V. Sahinidis. Derivative-free optimization: a review of algorithms and comparison of software implementations. *Journal of Global Optimization*, pages 1–47, July 2012. DOI: 10.1007/s10898-012-9951-y. 47

[149] R. T. Rockafellar. *Convex Analysis*. Princeton University Press, Princeton, New Jersey, 1970. 2, 40

[150] N. Satish, K. Ravindran, and K. Keutzer. A decomposition-based constraint optimization approach for statically scheduling task graphs with communication delays to multiprocessors. In *DATE '07*. DOI: 10.1145/1266366.1266381. 97, 98

[151] L. Schrage. *Optimization Modeling with LINGO*. 1999. 31

[152] A. Schrijver. *Theory of Linear and Integer Programming*. John Wiley & Sons, 1986. 2, 28, 40

[153] A. Schrijver. On the history of the transportation and maximum flow problems. *Mathematical Programming*, 91:437–445, 2002. DOI: 10.1007/s101070100259. 23

[154] A. Sen, H. Deng, and S. Guha. On a graph partitioning problem with applications to vlsi layout. In *Circuits and Systems, 1991., IEEE International Sympoisum on*, pages 2846 –2849 vol.5, June 1991. DOI: 10.1109/ISCAS.1991.176137. 3

[155] A. Shapiro, D. Dentcheva, and A. Ruszczyński. *LECTURES ON STOCHASTIC PRO-GRAMMING.* SIAM, 2009. DOI: 10.1137/1.9780898718751. 2, 14

[156] D. M. Shepard, M. C. Ferris, G. Olivera, and T. R. Mackie. Optimizing the Delivery of Radiation to Cancer Patients. *SIAM Review*, 41:721–744, 1999. DOI: 10.1137/S0036144598342032. 3

[157] A. Smith, G. Constantinides, and P. Cheung. Integrated floorplanning, module-selection, and architecture generation for reconfigurable devices. *Very Large Scale Integration (VLSI) Systems, IEEE Transactions on*, 16(6):733 –744, june 2008. DOI: 10.1109/TVLSI.2008.2000259. 3

[158] B. Speitkamp and M. Bichler. A mathematical programming approach for server con-solidation problems in virtualized data centers. *IEEE Transactions on Services Computing*, 3(4):266–278, 2010. DOI: 10.1109/TSC.2010.25. 73

[159] K. Srinivasan, K. Chatha, and G. Konjevod. Linear-programming-based techniques for synthesis of network-on-chip architectures. *Very Large Scale Integration (VLSI) Systems, IEEE Transactions on*, 14(4):407 –420, april 2006. DOI: 10.1109/TVLSI.2006.871762. 4

[160] S. Sutanthavibul, E. Shragowitz, and J. Rosen. An analytical approach to floorplan design and optimization. In *Design Automation Conference, 1990. Proceedings., 27th ACM/IEEE*, pages 187 –192, jun 1990. DOI: 10.1109/DAC.1990.114852. 3

[161] S. Swanson, K. Michelson, A. Schwerin, and M. Oskin. Wavescalar. In *Proceedings of the 36th annual IEEE/ACM International Symposium on Microarchitecture*, MICRO 36, pages 291–, 2003. DOI: 10.1109/MICRO.2003.1253203. 75

[162] M. Tawarmalani and N. V. Sahinidis. *Convexification and Global Optimization in Con-tinuous and Mixed-Integer Nonlinear Programming: Theory, Algorithms, Software and Appli-cations*, volume 69 of *Nonconvex Optimization and its Applications*. Kluwer Academic Pub-lishers, Dordrecht, 2002. 31

[163] M. Tawarmalani and N. V. Sahinidis. A polyhedral branch-and-cut approach to global optimization. *Mathematical Programming*, 103(2):225–249, 2005. DOI: 10.1007/s10107-005-0581-8. 31

[164] M. Tawarmalani and N. V. Sahinidis. A polyhedral branch-and-cut approach to global optimization. *Math. Program.*, 103(2):225–249, June 2005. DOI: 10.1007/s10107-005-0581-8. 110

[165] R. Thaik, N. Lek, and S.-M. Kang. A new global router using zero-one integer linear programming techniques for sea-of-gates and custom logic arrays. *Computer-Aided Design of Integrated Circuits and Systems, IEEE Transactions on*, 11(12):1479 –1494, dec 1992. DOI: 10.1109/43.180262. 3

[166] M. Thuresson, M. Sjalander, M. Bjork, L. Svensson, P. Larsson-Edefors, and P. Stenstrom. Flexcore: Utilizing exposed datapath control for efficient computing. In *IC-SAMOS 2007*. DOI: 10.1109/ICSAMOS.2007.4285729. 75

[167] Tomlab Optimization. http://tomopt.com/tomlab/. 47

[168] P. Van Hentenryck. *The OPL Optimization Programming Language*. MIT Press, 1999. 47

[169] L. Vandenberghe and S. Boyd. Semidefinite Programming. *SIAM Review*, 38(1):49–95, 1996. DOI: 10.1137/1038003. 14

[170] N. Vasić, D. Novaković, S. Miučin, D. Kostić, and R. Bianchini.   Dejavu: accelerating resource allocation in virtualized environments.   In *Proceedings of the seventeenth international conference on Architectural Support for Programming Languages and Operating Systems*, ASPLOS XVII, pages 423–436, New York, NY, USA, 2012. ACM. DOI: 10.1145/2150976.2151021. 62

[171] G. Venkatesh, J. Sampson, N. Goulding, S. Garcia, V. Bryksin, J. Lugo-Martinez, S. Swanson, and M. B. Taylor. Conservation cores: reducing the energy of mature computations. In *ASPLOS XV*. DOI: 10.1145/1736020.1736044. 75

[172] J. P. Vielma, S. Ahmed, and G. Nemhauser. Mixed-Integer Models for Nonseparable Piecewise-Linear Optimization: Unifying Framework and Extensions. *Operations Research*, 58(2):303–315, Oct. 2009. DOI: 10.1287/opre.1090.0721. 36, 39

[173] J. von Neumann and O. Morgenstern. *Theory of Games and Economic Behavior*, volume 2 of *Princeton Classic Editions*. Princeton University Press, 1944. 1

[174] H. M. Wagner. An integer linear-programming model for machine scheduling. *Naval Research Logistics Quarterly*, 6(2):131–140, 1959. DOI: 10.1002/nav.3800060205. 75, 97, 98

[175] E. Waingold, M. Taylor, D. Srikrishna, V. Sarkar, W. Lee, V. Lee, J. Kim, M. Frank, P. Finch, R. Barua, J. Babb, S. Amarasinghe, and A. Agarwal. Baring It All to Software: RAW Machines. *Computer*, 30(9):86–93, 1997. DOI: 10.1109/2.612254. 75

[176] Z. Wang and M. F. O'Boyle. Mapping parallelism to multi-cores: a machine learning based approach. In *PPoPP '09*, pages 75–84. DOI: 10.1145/1594835.1504189. 113

[177] M. Watkins, M. Cianchetti, and D. Albonesi. Shared reconfigurable architectures for cmps. In *FPGA 2008*, pages pages 299–304. DOI: 10.1109/FPL.2008.4629948. 75

[178] H. P. Williams. *Model Building in Mathematical Programming*. John Wiley & Sons, third edition, 1990. 38

[179] L. A. Wolsey. *Integer Programming.* John Wiley & Sons, New York, NY, 1998. 2

[180] T.-H. Wu, A. Davoodi, and J. Linderoth. Grip: Scalable 3d global routing using integer programming. In *Design Automation Conference, 2009. DAC '09. 46th ACM/IEEE*, pages 320 –325, july 2009. DOI: 10.1145/1629911.1629999. 3

[181] J. Xu, M. Zhao, J. Fortes, R. Carpenter, and M. Yousif. Autonomic resource management in virtualized data centers using fuzzy logic-based approaches. *Cluster Computing*, 11(3):213–227, Sept. 2008. DOI: 10.1002/9780470382776. 72

[182] T. C. Xu, P. Liljeberg, and H. Tenhunen. Optimal memory controller placement for chip multiprocessor. In *CODES+ISSS*, pages 217–226, 2011. DOI: 10.1145/2039370.2039405. 113

[183] P. Yu and T. Mitra. Scalable custom instructions identification for instruction-set extensible processors. In *Proceedings of the 2004 international conference on Compilers, architecture, and synthesis for embedded systems*, CASES '04, pages 69–78, 2004. DOI: 10.1145/1023833.1023844. 60

[184] W. Zhou, Y. Zhang, and Z. Mao. Pareto based multi-objective mapping ip cores onto noc architectures. In *IEEE Asia Pacific Conference on Circuits and Systems, 2006.*, pages 331 –334. DOI: 10.1109/APCCAS.2006.342418. 113

[185] X. Zhu, C. Santos, D. Beyer, J. Ward, and S. Singhal. Automated application component placement in data centers using mathematical programming. *International Journal of Network Management*, 18(6):467–483, 2008. DOI: 10.1002/nem.707. 73

Authors' Biographies

TONY NOWATZKI

Tony Nowatzki is a graduate student at the University of Wisconsin-Madison, working as a research assistant in the Vertical Research Group. His research centers around computational accelerators from a design exploration and comparison perspective. Broad interests include architecture and compiler co-design. He is a student member of IEEE. He has a Bachelor's of Computer Science and Computer Engineering from the University of Minnesota, and a Master's of Computer Science from UW-Madison.

MICHAEL FERRIS

Michael Ferris is a Professor at the University of Wisconsin-Madison in the department of computer sciences. His research is concerned with algorithmic and interface development for large scale problems in mathematical programming, including links to the GAMS and AMPL modeling languages, and general purpose software such as PATH, NLPEC, and EMP. He has worked on several applications of both optimization and complementarity, including cancer treatment plan development, radiation therapy, video-on-demand data delivery, economic and traffic equilibria, structural and mechanical engineering. Ferris is a SIAM fellow and an INFORMS fellow and received the Beale-Orchard-Hays prize from the Mathematical Programming Society and is a past recipient of a NSF Presidential Young Investigator Award, and a Guggenheim Fellowship. He serves on the editorial boards of Mathematical Programming, SIAM Journal on Optimization, Transactions of Mathematical Software, and Optimization Methods and Software.

KARTHIKEYAN SANKARALINGAM

Karthikeyan Sankaralingam is an Associate Professor at the University of Wisconsin-Madison in the department of computer sciences. He leads the Vertical Research group at UW-Madison, exploring a vertically integrated approach to microprocessor design. His research has developed widely cited results on Dark Silicon, hardware specialization in the DySER architecture, and novel generalizations of GPUs. He is a recipient of the NSF Career Award in 2009 and the IEEE TCCA Young Computer Architect Award in 2011. He is an IEEE Senior Member. He got his PhD and MS from the University of Texas at Austin, and his Bachelor's degree from the Indian Institute of Technology, Madras.

CRISTIAN ESTAN

Cristian Estan is an architect at Broadcom Corporation where he works on coprocessors performing critical tasks for networking infrastructure such as packet classification, forwarding lookups and deep packet inspection. He has achieved major reductions in power consumption and cost per bit and increases in functionality through algorithmic and architectural innovation. He has received the Broadcom CEO achievement recognition award (2013), PLDI distinguished paper award (2013), NSF CAREER award (2006), ACSAC best paper award (2006) and UCSD CSE PhD dissertation award (2004). Earlier he worked at NetLogic Microsystems, taught at the CS Department of University of Wisconsin-Madison and had shorter stints at various startups. He published 30 research papers at selective peer-reviewed venues in the fields of computer networking, security, systems, programming languages and databases and is an inventor on 14 patents and patent applications.

NILAY VAISH

Nilay Vaish is a graduate student at the University of Wisconsin-Madison, working as a research assistant in the Multifacet Group. He has a Bachelor's degree from the Indian Institute of Technology, Delhi and a Master's degree from UW-Madison.

DAVID WOOD

David Wood is a Professor and Romnes Fellow in the Computer Sciences Department at the University of Wisconsin, Madison. He also holds a courtesy appointment in the Department of Electrical and Computer Engineering. He received a B.S. in Electrical Engineering and Computer Science (1981) and a Ph.D. in Computer Science (1990), both at the University of California, Berkeley. He joined the faculty at the University of Wisconsin in 1990. Dr. Wood was named an ACM Fellow (2005) and IEEE Fellow (2004), received the University of Wisconsin's H.I. Romnes Faculty Fellowship (1999), and received the National Science Foundation's Presidential Young Investigator award (1991). Dr. Wood is Area Editor (Computer Systems) of ACM Transactions on Modeling and Computer Simulation, is Associate Editor of ACM Transactions on Architecture and Compiler Optimization, served as Program Committee Chairman of ASPLOS-X (2002), and has served on numerous program committees. Dr. Wood is an ACM Fellow, an IEEE Fellow, and a member of the IEEE Computer Society. Dr. Wood has published over 70 technical papers and is an inventor on thirteen U.S. and International patents.